ALL THE WORLD'S ANIMALS
AQUATIC INVERTEBRATES

ALL THE WORLD'S ANIMALS
AQUATIC INVERTEBRATES

TORSTAR BOOKS
New York · Toronto

CONTRIBUTORS

GJB   Gerald J. Bakus PhD
University of Southern
California
Los Angeles, California
USA

RB   Robin C. Brace BSc PhD
University of Nottingham
England

AC   Andrew Campbell BSc DPhil
Queen Mary College
University of London
England

JEC   June E. Chatfield BSc PhD
ARCS
Gilbert White Museum
Selborne, Hants
England

GD   Gordon Dickerson BSc MIBiol
Formerly of Wellcome
Research Laboratory
Beckenham, Kent
England

PRG   Peter R. Garwood BSc PhD
Dove Marine Laboratory
University of Newcastle upon
Tyne
England

JL-P   Johanna Laybourn-Parry BSc
MSc PhD
University of Lancaster
England

PSR   Philip S. Rainbow MA PhD
Queen Mary College
University of London
England

ALL THE WORLD'S ANIMALS
AQUATIC INVERTEBRATES

TORSTAR BOOKS
300 E. 42nd Street,
New York, NY 10017

Project Editor: Graham Bateman
Editors: Bill MacKeith, Robert Peberdy
Art Editor: Jerry Burman
Art Assistant: Carol Wells
Picture Research: Alison Renney
Production: Clive Sparling
Design: Chris Munday, Andrew Lawson
Index: Barbara James

Originally planned and produced by:
Equinox (Oxford) Ltd
Littlegate House
St Ebbe's Street
Oxford OX1 1SQ

Editors
Dr Keith Bannister
British Museum
(Natural History)
London
England

Dr Andrew Campbell
Queen Mary College
University of London
England

Advisory Editors
Professor Fu-Shiang Chia
University of Alberta
Edmonton
Canada

Dr John E. McCosker
Steinhart Aquarium
California Academy of
Sciences
San Francisco
USA

Artwork panels and diagrams
S. S. Driver
Richard Lewington
Richard Gorringe

Mick Loates
Barbara Cooper

On the cover: Sea slug
Page 1: Jellyfish
Pages 2–3: Ghost crab
Pages 4–5: Sea anemone
Pages 6–7: Banded coral shrimp
Pages 8–9: Starfish

10 9 8 7 6 5 4 3 2 1

Printed in Belgium

Library of Congress Cataloging in Publication Data

Aquatic invertebrates.

  (All the world's animals)
  Bibliography: p.
  Includes index
  1. Aquatic invertebrates. I. Series.
QL120.A68   1987   592.092   86–11332

ISBN 0-920269-72-9 (Series: All the World's Animals)
ISBN 0-920269-85-0 (Aquatic Invertebrates)

In conjunction with *All the World's Animals*
Torstar Books offers a 12-inch raised
relief world globe.
  For more information write to:
  Torstar Books
  300 E. 42nd Street
  New York, NY 10017

CONTENTS

FOREWORD

Hundreds of millions of years of evolution have given birth to a bewildering array of aquatic invertebrates, from the single-celled amoeba to the giant squid, from the delicate anemones, corals and sponges to the tough-shelled crustaceans.

Aquatic invertebrates are animals without backbones that live in the sea, fresh water or a moist terrestrial environment. Among their number are many parasites, such as the malarial parasite *Plasmodium*, whose "aquatic environment" is that of the bodies of their hosts. And while many of these animals remain unknown to the layman there is no reason to think the aquatic invertebrates are the poor relations of the more familiar vertebrates.

For sheer beauty, the microscopic architecture of the heliozoans or sun animalcules and the sea anemones is not easily surpassed. For complexity of structure, intelligence and behavior, the squids and octopuses come near to the fish in their mastery of the sea. And for enigmatic architecture and diverse solutions to evolutionary problems, the starfishes, sea urchins and their allies are preeminent in the animal world.

The latest techniques in underwater and microscopic photography provide *Aquatic Invertebrates* with a stunning backcloth of colors, astonishing shapes and impressive waterscapes. Expanding the rich and detailed text are color artwork panels depicting animal forms and ecological scenes, and simple line diagrams revealing the inside stories of the invertebrates' anatomy, life-cycles or adaptations to their environment.

How this book is organized

The aquatic invertebrates contain an immense variety of often small and inconspicuous animals whose diversity reflects their long evolutionary history. More than 30 major subdivisions ("phyla"—singular "phylum") of invertebrates are recognized by scientists and nearly all are covered in this volume. The only invertebrates not covered are the terrestrial arthropods—the insects, millipedes and their relatives (phylum Uniramia), the spiders and other land-dwelling arachnids (most of the phylum Chelicerata), and the velvet worms (phylum Onychophora). Even the phylum Chordata, which includes vertebrates such as fish, mammals, birds, reptiles and amphibians, has some invertebrate members, such as sea squirts and lancelets, and these are covered here.

Some invertebrate phyla, while overwhelmingly aquatic in habit, contain groups that have terrestrial forms. For example, although segmented worms are mainly marine, the earthworms live in damp soil; slugs and snails are terrestrial variations of the mainly aquatic mollusks. For completeness, such terrestrial forms are included here. The diversities of form and biology are huge—a salmon and an elephant have more in common with each other than do many apparently related members of the invertebrate phyla. Thus the subject is enormous. Despite the wealth of evolutionary enterprise and extensive distribution of the invertebrates, most people know very little about them. For this reason it is very difficult to refer to many of these animals without scientific terminology.

A simplification of the extensive classification of invertebrates is needed. To achieve this, the organisms treated here have been classified into a few subjective groupings which are: Simple Invertebrates, Sedentary and Free-swimming Invertebrates, Worm-like Invertebrates, Jointed-limbed Invertebrates, and Mollusks, Echinoderms and Sea Squirts. Each of these major headings is introduced by an essay dealing with the ecological attributes which members of the group have in common, and indicating relationships. Then each phylum is dealt with, treating the classification, structure, development and ecology as appropriate.

Each major entry starts with a fact panel which summarizes the biology of the group concerned and gives the scientific names of those few species that have valid common names. This work is not structured at a species level because the species it covers are far too numerous to be treated in this way. In cases of large and conspicuous phyla like the Protozoa, Annelida, Crustacea and Mollusca, the main subdivisions of the phylum (often the classes) are dealt with in a classificatory order. In the smaller groups a general résumé of the whole is given. A number of topics worthy of special attention are drawn out as boxed features, or in the case of some ecologically, economically or medically important subjects, as full double-page articles.

Every phylum has its own special form, parts of which may be peculiar to it and have unfamiliar names. A glossary and detailed illustrations will enable the lay reader to understand the scientific terms used and to appreciate the complex internal structures lying beneath apparently simple exteriors.

An international team of specialists have contributed to *Aquatic Invertebrates*, each dealing with his or her own special area from an individual standpoint. In this way particular impressions of the phyla are built up and valid zoological opinions developed. We have allowed these to speak for themselves, since the development of ideas and the criticism of theories is the very essence of zoological progress.

AQUATIC INVERTEBRATES

KINGDOM PROTISTA—SINGLE-CELLED ANIMALS

Subkingdom Protozoa

Flagellates, amoebae, opalinids
Phylum: Sarcomastigophora

Ciliates
Phylum: Ciliophora

Phylum: Apicomplexa—includes malaria, coccidiosis

Phylum: Labyrinthomorpha

Phylum: Microspora

Phylum: Ascetospora

Phylum: Myxospora

KINGDOM: ANIMALIA

Subkingdom Parazoa

Sponges
Phylum: Porifera

Subkingdom Mesozoa

Mesozoans
Phylum: Mesozoa

Subkingdom Metazoa—Multi-cellular Animals

Sea anemones, jellyfishes, corals and allies
Phylum: Cnidaria

Comb jellies or sea gooseberries
Phylum: Ctenophora

Endoprocts
Phylum: Endoprocta (Entoprocta or Kamptozoa)

Rotifers or wheel animalcules
Phylum: Rotifera

Kinorhynchs
Phylum: Kinorhyncha

Gastrotrichs
Phylum: Gastrotricha

Lampshells
Phylum: Brachiopoda

Moss animals or sea mats
Phylum: Bryozoa (Ectoprocta)

Horseshoe worms
Phylum: Phoronida

Flatworms (flukes, tapeworms)
Phylum: Platyhelminthes

Ribbon worms
Phylum: Nemertea

Roundworms
Phylum: Nematoda

Spiny-headed or thorny-headed worms
Phylum: Acanthocephala

Horsehair worms or threadworms
Phylum: Nematomorpha

Segmented worms
Phylum: Annelida

Echiurans
Phylum: Echiura

Priapulans
Phylum: Priapula

Sipunculans
Phylum: Sipuncula

Beard worms
Phylum: Pogonophora

Arrow worms
Phylum: Chaetognatha

Acorn worms and allies
Phylum: Hemichordata

Crustaceans
Phylum: Crustacea

Chelicerates
Phylum: Chelicerata

Uniramians
Phylum: Uniramia
Terrestrial, including insects, millipedes, spiders, scorpions.

Water bears
Phylum: Tardigrada

Tongue worms
Phylum: Pentastomida

Velvet or walking worms
Phylum: Onychophora
Terrestrial.

Mollusks
Phylum: Mollusca

Spiny-skinned animals
Phylum: Echinodermata

Chordates
Phylum: Chordata
Includes the aquatic sea squirts and lancelets, and also all vertebrates.

WHAT do crabs, sea urchins, earthworms, malaria parasites and corals have in common? In fact, these very diverse groups share very little, apart from the fact that they all lack a backbone. Of the 1,071,000 or so known species of animal about 1,029,300 are invertebrate, that is, over 95 percent are animals without backbones. Invertebrates make up the bulk of animals, measured both in terms of numbers of species recognized and numbers of individuals. Some invertebrates, like garden snails and earthworms, are conspicuous and familiar animals, while others, although abundant, pass unnoticed by most people.

Invertebrate body forms range in size from the lowly microscopic *Amoeba*, which may be just one micrometer in diameter, to the Giant squid 59ft (18m) in length, a ratio of 1:18,000,000. They include life forms as diverse as the Desert locust and the sea anemones. They inhabit all regions of the globe, and all habitats, from the ocean abyss to the air. Life almost certainly originated in the seas, and virtually all the major invertebrate groups (phyla) have marine representatives. Somewhat fewer (almost 14 phyla) have conquered freshwater. Fewer still (about 5 phyla) live on land and of these only the jointed-limbed groups (arthropods) have mastered the air and really dry places. Most numerous among arthropods are the uniramians (eg insects, millipedes, centipedes) and the chelicerates (eg scorpions, spiders, ticks, mites).

Many invertebrates, like slugs (whether of the garden or sea), are free living; others, such as barnacles, are attached to the substrate throughout their adult life; yet others live as parasites in or on the bodies of plants or other animals. Some invertebrates are of great commercial significance, either as direct food for man (eg prawns and oysters), or as food for man's exploitable reserves (eg the planktonic copepods on which herring feed). Others (eg earthworms) are much appreciated because they improve the soil for agriculture. Many invertebrates live as parasites, either in man himself or in the bodies of domestic animals and plants, where great damage may be wrought, and so are of great medical or agricultural importance.

This great diversity of form and life-style in animals has led zoologists to sort out, or classify, animals according to type and evolutionary connections. To be certain that they are speaking of the same animals, they give each species a unique scientific name (eg *Lumbricus terrestris*, the Common earthworm). Every species is classified into one of the major groups, or phyla. A phylum comprises all those animals which are thought to have a common evolutionary origin. According to their understanding of the probable evolutionary processes involved, zoologists may recognize some 39 animal phyla. With one exception, they are made up exclusively of invertebrate animals. The phylum Chordata includes all animals with a hollow dorsal nerve cord. Nearly all—including fishes, amphibians, reptiles, birds—have a backbone, but some are invertebrate, such as the sea squirts.

The technical classification that zoologists employ leads from the most primitive and simple animal types to the most complex and advanced. In order to achieve some form of system, various levels of organization are recognized which give clear distinctions between phyla. The most fundamental of these lies in the number of cells in the body. A cell is the smallest functional unit

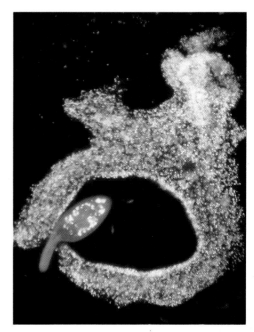

▲ **Cell eats cell.** A microscopic *Amoeba* engulfs a ciliate *Paramecium*, both single-celled animals (× 100).

◄ **Sophisticated cephalopod.** The octopus is the most intelligent of all invertebrate animals. It has a memory and can learn.

Classification of a species, *Asterias rubens*, the European Common starfish.

Kingdom Animalia
Subkingdom Metazoa
Phylum Echinodermata
Subphylum Asterozoa
Class Asteroidea
Subclass Euasteroidea
Order Forcipulatida
Family Asteriidae
Genus *Asterias*
Species *rubens*

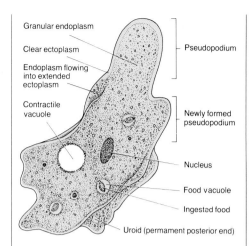

Granular endoplasm

Clear ectoplasm

Endoplasm flowing
into extended
ectoplasm

Contractile
vacuole

Pseudopodium

Newly formed
pseudopodium

Nucleus

Food vacuole

Ingested food

Uroid (permanent posterior end)

▲ **The single protozoan cell** may be quite a complex structure, with specialized parts called organelles responsible for different functions including feeding and locomotion (× 125).

▼ **Number and arrangement of cell layers** distinguish degrees of complexity in many-celled animals, from the single layer of monoblastic sponges and mesozoans (1), through the diploblastic jellyfishes and comb jellies (2), to the three layers of most animals. The bulky middle layer (mesoderm) of triploblastic animals may be solid, as in flatworms and ribbon worms (3), or divided into inner and outer parts separated by a cavity or coelom (4). Some groups (eg nematode worms) have a body cavity that is not formed within the mesoderm and is called a pseudocoel (5).

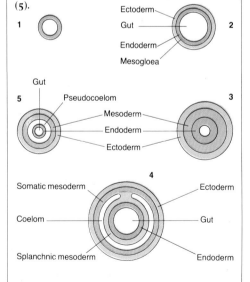

1

Ectoderm
Gut
Endoderm
Mesogloea
2

5
Gut
Pseudocoelom
Mesoderm
Endoderm
Ectoderm
3

4
Somatic mesoderm
Coelom
Splanchnic mesoderm
Ectoderm
Gut
Endoderm

▶ **Relationships between animal phyla,** showing grade of organization, numbers of cell layers, presence or absence of a coelom, and site of development of the mouth in the embryo—the protostome/deuterostome distinction.

of an animal, governed by its own nucleus which contains genetic material, known as DNA.

Protozoans—the "simplest animals"

The animals with bodies made up of a single cell represent a separate level of organization from all the rest, for their body processes are performed by the one cell, which therefore itself cannot be specialized. In multicellular animals the responsibility for different life processes is shared out, different cells being specialized for different tasks, eg receiving stimuli (sensory cells), communication (nerve cells), movement (muscle cells) etc.

The single-celled animals or protozoans (sometimes referred to as acellular, ie without distinct cells) are assumed to be low on the scale of evolutionary sophistication. In reality their single cell is often large and complex. The essential life processes are carried out by special regions (organelles) within the cell—nucleus for government, bubble-like food vacuoles for energy acquisition, vacuoles that contract and expand to regulate water levels within the cell and so on.

Another fundamental question may be asked about single-celled animals. Are they in fact all animals? True animals derive their energy from relatively complex organic material that is plant or animal in origin—they are heterotrophic; they obtain their nourishment by a process of breaking down organic materials (are catabolic). However, some protozoans (eg green flagellates) have developed structures characteristic of plants, called chloroplasts, which are incorporated into their cells. The chloroplast enables the cell to synthesize organic materials, such as sugars, from mineral salts, carbon dioxide and water in the presence of sunlight (photosynthesis). Like green plants, these protozoans can subsist on inorganic food (are autotrophic) from which they synthesize organic material (are anabolic). Botanists consider such protozoans to be single-celled plants. The difficulty is that while some protozoans are clearly plants and others clearly animals, some, like *Euglena gracilis*, can feed by both autotrophy *and* heterotrophy. For reasons such as this some authorities believe that most protozoans warrant classification as an animal subkingdom, with obviously

			PROTOSTOME	Protostome with some deuterostome tendencies	DEUTEROSTOME	
Organ grade	with coelom and blood vascular system	**Metameric** (many segments)	Arthropoda (7 phyla) Annelida		Chordata	Subkingdom **Metazoa**
		Oligomeric (few segments)	Pogonophora Mollusca	Brachiopoda Ectoprocta Phoronida Chaetognatha	Hemichordata Echinodermata	
		Americ (lacking segments)	Echiura Sipuncula Priapula			
	lacking true coelom	**Pseudocoelomate** vascular	Nemertea			
		Pseudocoelomate non-vascular		Endoprocta Acanthocephala Nematoda Nematomorpha Gastrotricha Rotifera Kinorhyncha		
		Acoelomate (no coelom)	Platyhelminthes			
		Triploblastic (3 layers)				
Tissue grade		**Diploblastic** (2 layers)	Ctenophora Cnidaria			
Cellular grade		**Monoblastic** (1 layer)	Porifera Mesozoa			Subkingdom **Parazoa** Subkingdom **Mesozoa**
		Multi-cellular				**KINGDOM ANIMALIA**
		Single-celled	Protozoa (7 phyla)			**KINGDOM PROTISTA**

plant-like protozoans treated as members of the plant kingdom, outside it.

Recently scientists have proposed a new classification of the living world which overcomes the problem of whether protozoans are plants or animals. This reorganization proposes five kingdoms: Monera (bacteria and blue-green algae); Protista; Plantae (plants); Fungi and Animalia. The protozoans are included in the Protista as the sub-kingdom Protozoa, along with some other groups previously classified as algae by botanists.

Origins of life

Protozoans are most important because, as the "simplest animals" they are likely to provide important keys to two fundamental questions. These concern the origin of life, and the origin of multi-cellular metazoan animals. The origin of life is shrouded in scientific speculation. The biblical account of the origin of life as presented in *Genesis* is an historic attempt to answer one of man's most fundamental questions. Belief in the idea of a Divine creation is a matter of faith which cannot be tested by science. Nor can science yet tell us how life first began, although it has also been shown that primitive ideas such as spontaneous generation of life are completely wrong.

It is thought that the earth is not quite 5,000 million years old, and realistic estimates suggest that life began in its simplest form 4,000 million years ago. The first sedimentary rocks, not quite 4,000 million years old, contain fossils of simple cells which resemble those of present-day bacteria, that is, they lacked a distinct nucleus (ie were prokaryotes). These lived in a primeval atmosphere devoid of oxygen. The appearance of oxygen on the earth, 1,800 million years ago, brought with it many new evolutionary possibilities. The protozoans, green algae, higher plants and animals appeared a lot later, the earliest known fossils having been taken from rocks 1,000 million years old. All these organisms are made up of cells with nuclei (ie are eukaryotes). Thus the startling fact emerges that for three-quarters of the period for which life has existed on earth the only cells were prokaryotes, ie resembling bacteria. It was not until the Cambrian period 600–500 million years ago that invertebrates such as mollusks, trilobites, lampshells and echinoderms became established. Invertebrates with soft delicate bodies, such as flatworms and sea squirts, have not left any fossil record.

For life to have appeared many conditions

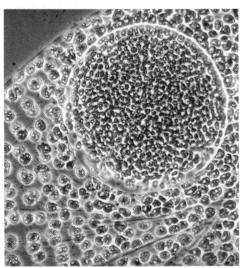

▲ **Colony of polyps.** Cnidarian colonies like this gorgonian arise from a many-celled floating planula larva which settles, develops into a polyp, and buds off other polyps asexually. The form of the colony varies with the species and situation.

◄ **Sharing common ancestry** with the higher animals. A colony of single-celled flagellates (*Volvox* species), with a daughter colony in formation (× 200). Ancestors of these colonies may have developed into permanent two-layered organisms somewhat resembling the planula larvae of sea anemones, corals and jellyfishes.

had to be fulfilled. It seems quite possible that the physical conditions prevailing on the surface of the early earth could have generated simple organic molecules such as amino acids, and then proteins, from inorganic molecules. The big unanswered question is how such substances could form themselves into organized living systems capable of reproducing their own kind.

Multi-cellular animals

The origins of multi-cellular animals are also speculative, but rather more can usefully be said about their possible early history. Because so many of the early animals had soft bodies they left very little fossil record. All theories about the early evolution of animals therefore rely mainly on the study of similarities between developing embryos and adults of animals in different groups. This allows inferences to be drawn about common ancestry. Two chief theories have been put forward.

According to one theory, protozoan animals gave rise to multi-cellular ones by colony formation. A number of types of colonial protozoan are known to exist, such as *Volvox*.

The famous 19th-century German biologist Haeckel proposed that a hollow *Volvox*-like ancestor could have developed into a two-layered organism. Views differ as to whether or not this was a planktonic or a bottom-dwelling organism, but it may have somewhat resembled the planula larvae of the sea anemones and jelly fishes (p13) and could have given rise to bottom-dwelling animals such as adult hydroids and sea anemones. Although most animals are composed of three layers of cells, often in a highly modified form, there is evidence that the evolution from two-layered animals did occur in evolutionary history.

A different theory proposes that multi-cellular animals arose from single-celled animals containing many nuclei by the growth of cell walls between each nucleus. A number of protozoans, for example *Opalina*, are like this (see p22). According to this theory, the primitive multi-cellular animal would lack a gut, as in present-day gutless flatworms (see p62). However, the latter have three layers of cells, which raises the question—what is the origin of the two-layered animals? Because of this and other criticisms, this theory is now generally discarded in favor of the colonial one.

While it is not certain how the multi-cellular animals evolved from protozoan ancestors, it is possible to distinguish groups of metazoans on the basis of relative sim-

plicity or complexity of structure. (Some biologists divide invertebrates into protostomes and deuterostomes, see below.)

In metazoans there are three categories that can be used to determine level of complexity: how the cells are organized; how many layers of cells are to be found within the body; and whether or not a body cavity is present.

There are few types of cell in the most lowly metazoans and in sponges and these cells are never arranged into groups of similar cells (tissues). Such animals are said to have a *cellular grade of organization*. In the next step, as found in jellyfishes and allies and comb jellies, cells with similar functions are arranged together into tissues, each tissue having its own function or series of functions—these animals have a *tissue grade of organization*. In all animals apart from those just mentioned, requirements for functional specialization increase such that specific organs (often comprising a series of tissues) have evolved. Thus all animals from flatworms to man are said to have an *organ grade of organization*.

The second way to divide up multi-cellular animals is to look at the number of layers of cells that make up the animal's body. Mesozoans and sponges consist of just one layer of cells, but in the jellyfishes and comb jellies two layers appear (ectoderm outside and endoderm inside). This "diploblastic" condition contrasts markedly with the single layer of cells seen in the protozoan colony *Volvox* (p22), which is described as monoblastic. The two layers develop from the egg and remain throughout adult life, separated from each other by a sheet of jelly-like mesogloea.

▲ **Drifting predator,** the jellyfish *Rhizostoma octopus* of northern waters. In jellyfishes the two cell layers of the body are separated by an extensive jelly-like mesogloea. Jellyfishes lack a brain and their movement is under control of a primitive nerve net. Most jellyfishes live in warm waters.

▶ **Crustacean in a mollusk shell.** A hermit crab (*Dardanus lagopodes*) takes shelter in the former home of a shellfish. Eyes, antennae, palps, pincers, external gills and legs are paired appendages on each body segment, evolved to fulfill a wide range of functions, including sense perception, feeding, locomotion and copulation.

▼ **Crowns of fanworms** (here *Protula maxima*) act as feeding organs and gills. Giant nerve fibers running the length of these polychaete worms enable them to retract the gaily colored crown with startling rapidity.

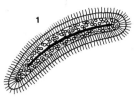

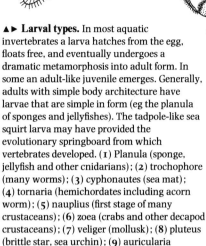

All animals "above" the jellyfishes and comb jellies are equipped with three layers of cells and are known as triploblastic. Here the ectoderm and endoderm are separated by a third cell layer, the mesoderm. The mesoderm forms the most bulky part of many animals, contributing the musculature of the body wall as well as that of the gut. In some of these triloblastic animals, eg flatworms (Platyhelminthes) and ribbon worms (Nemertea), the mesoderm is solid and not itself divided into two layers by a body cavity (coelom), so they are described as acoelomate. Without a body cavity these animals are at a disadvantage because the body movements affect the movements of the gut and vice versa. To be able to move the gut independently of the body wall is a great advantage, as it allows for sophisticated digestive activities. Such a condition is reached only in coelomate animals, in which the mesoderm is divided into an outer section forming the body wall and an inner section forming the muscles of the gut, and the two are largely separated by a body cavity (coelom) within the mesoderm. The possession of a fluid-filled body cavity is the hallmark of all the more advanced invertebrates and the major phyla, including mollusks, annelids, the different arthropods, echinoderms and chordates, all have a coelom and are coelomates.

In some groups there is a body cavity between the body wall and the gut, but this is a pseudocoelom, not a true coelom, for it is not formed inside the mesoderm. Generally it persists from an early stage in development and it contains fluid. In animals such as the nematode worms it performs an important function as a fluid skeleton.

Some evolutionary biologists suggest a different way of grouping invertebrates, believing that two main evolutionary lines have emerged in the animal world, the protostomes (first mouth) and deuterostomes (second mouth). In the early embryos of protostomes such as annelid worms and insects, the mouth is formed at or near the site of the blastopore. In deuterostomes such as the echinoderms (starfishes and sea urchins) and the chordates, the anus forms at the blastopore. Other distinctions can be seen by comparing the development of the two types. In the annelids the nerve cord is on the underside, a double, solid structure reminiscent of the nerve cord of insects. In the chordates it is a single hollow structure along the upper side of the animal. The annelid coelom is formed by the splitting of the mesoderm, but in the echinoderms and the chordates it develops from two pouches of the primitive gut. The sea-dwelling representatives of these two "lines" have characteristically different larvae.

These distinctions are not definite evidence of links between phyla, but indications of possible evolutionary affinities. Certainly the chordates are likely to have arisen from a deuterostome type of ancestor, and the form of development shown by annelids and mollusks places them a long way from the echinoderms and chordates in any phylogeny or "family tree."

Symmetry

One of the most obvious differences separating the phyla of the animal kingdom is the overall appearance of the animals. The majority of animals, including the vertebrates, worms and jointed-limbed forms, are bilaterally symmetrical: complementary right and left halves are mirror images.

▲▶ **Larval types.** In most aquatic invertebrates a larva hatches from the egg, floats free, and eventually undergoes a dramatic metamorphosis into adult form. In some an adult-like juvenile emerges. Generally, adults with simple body architecture have larvae that are simple in form (eg the planula of sponges and jellyfishes). The tadpole-like sea squirt larva may have provided the evolutionary springboard from which vertebrates developed. (1) Planula (sponge, jellyfish and other cnidarians); (2) trochophore (many worms); (3) cyphonautes (sea mat); (4) tornaria (hemichordates including acorn worm); (5) nauplius (first stage of many crustaceans); (6) zoea (crabs and other decapod crustaceans); (7) veliger (mollusk); (8) pluteus (brittle star, sea urchin); (9) auricularia (starfish, sea cucumber); (10) "tadpole" larva of sea squirt.

◀▼ **Most animals are symmetrical.** The majority are bilaterally symmetrical with the left mirroring the right side, and with front and rear ends. In many worms a head is barely distinguishable, but jointed-limbed animals such as crustaceans, and most mollusks, have clearly developed heads.

The straightforward symmetry of the Banana slug (*Ariolimax californicus*) BELOW with its paired tentacles conceals the internal twisting of body organs common to most gastropod mollusks.

Two major groups show a different, radial symmetry without a head—the jellyfishes and their relatives, and echinoderms such as this starfish *Marthasterias glacialis* ABOVE LEFT with its five-rayed symmetry. One arm is in the process of regeneration, its predecessor having been broken or torn off by a predator.

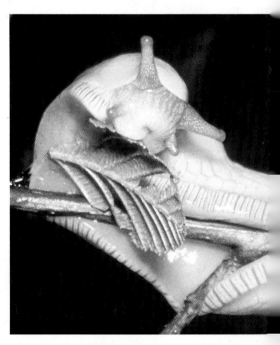

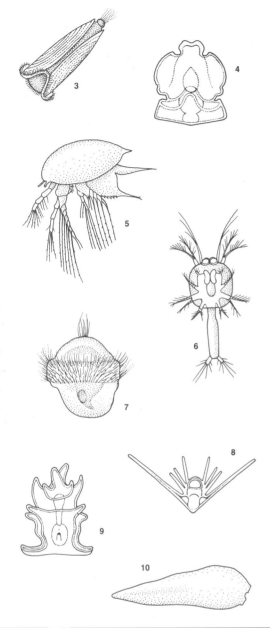

Their front and hind ends are however dissimilar. This is because in these animals there has been some specialization at the head end. In the lowly flatworms the head is only feebly developed, but it is identifiable. In the jointed-limbed animals, such as the insects, it is clearly defined. The evolution of a head end (cephalization) has come about in direct response to the development of forward movement. Clearly it is an advantage to have specialized sensory equipment (eg eyes, smelling and tasting receptors at the front of the body) to deal efficiently with environmental stimuli as they occur in the direction of travel. In this way prey and predators can be quickly identified.

The position of sense receptors is often associated with the mouth opening. These factors have led to the development of aggregations of nervous tissue to integrate the messages coming from the receptors and to initiate a coordinated response (eg attack or flight) in the muscle systems of the body. This brain, be it ever so simple, came to lie at the front end of the body, often near the mouth, and the typical head arrangement was formed. In addition to a distinct head, other features of bilaterally symmetrical animals are thoracic regions (modified for respiration) and abdominal regions (modified for digestion, absorption and reproduction). The function of locomotion may be undertaken by either or both of the regions. In the mollusks the principles of bilateral symmetry are somewhat disguised by the unique style of body architecture, but if these animals are reduced to their simplest form, as in the aplacophorans and chitons, a basic bilateral symmetry is visible.

Some animals have, however, managed without a head. Despite their relatively high position in the table of phyla, the echinoderms (eg starfishes, sea urchins) are headless. This is all the more surprising when one realizes that the earliest echinoderms and their likely ancestors were bilateral animals. Still, they managed without a head, and in the present-day echinoderms the nervous tisue is spread fairly evenly throughout the body and all sensory structures are very simple and widely distributed. Modern adult echinoderms are fundamentally radially symmetrical; the body parts are equally arranged around a median vertical axis which passes through the mouth. There is no clear front end, nor left and right sides, in most of the species. Radial symmetry is best for a stationary lifestyle, in which food is collected by nets or fans of tentacles. This was almost certainly the life-style of the earliest echinoderms, as it is of most of present-day Cnidaria (jellyfishes and their allies), the other phylum to display a clear radial symmetry. The radial symmetry in the Cnidaria is evident, whereas in the echinoderms a unique five-sided body form has been imposed on it, as can be seen in present-day starfishes and brittle stars.

Finally, there are those animals which are essentially without regular form or are asymmetrical. These are the sponges, which grow in a variety of fashions, including encrusting, upright or plant-like, or even boring into rocks. While each species has a characteristic development of the canal system inside the body, its precise external appearance depends very much on prevailing local conditions, such as exposure, currents and form of substrate. The simple body form of the sponges of course rules out the development of nerves and muscles, so a coordinating nerve system has not evolved.

The bilaterally symmetrical bodies of coelomate metazoans may be unsegmented (americ), divided into a few segments (oligomeric), or divided into many structural units (metameric). The bodies of many types (eg ectoprocts, or moss animals, and echinoderms) show division into a few segments during their development, but these may be masked in the adult. In the annelids and arthropods the repetition of structural units along the body (metamerism) allows for the modification of the segmental appendages to fulfill various functions. In the arthropods these appendages or limbs, often jointed, carry out a wide range of activities, from locomotion to copulation.

AC

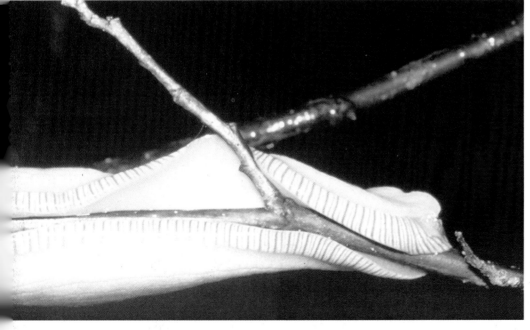

Drifters and Wanderers

The ecology of zooplankton

The word "plankton" means drifter or wanderer. It refers to those plants and animals that are swept along by water currents rather than by their own swimming ability. (Animals that swim and determine their own direction are called nekton.) The greatest diversity of plankton exists in the world's seas and oceans, but lakes and some rivers have their own plankton communities. Here examples from the sea will be used. Plants of the plankton are known as phytoplankton and animals of the plankton as zooplankton. While many zooplankton are small animals, less than 0.2in (5mm) long, a few are large, for example some jellyfishes, whose tentacles may reach 49ft (15m) in length or more. A number of zooplankton can actually swim, but not sufficiently well to prevent them from being swept along by currents in the water. However their swimming ability may be sufficient to allow them to regulate their vertical position in the water which can be very important as the position or depth of their food can vary around the daytime/nighttime cycle (for example phytoplankton rises by day and sinks by night whereas zooplankton does the reverse).

Seawater contains many nutrients important for plant growth, notably nitrogen, phosphorus and potassium. Their presence means that phytoplankton can photosynthesize and grow while they drift in the illuminated layers of the sea. Two forms of phytoplankton, dinoflagellates and diatoms, are particularly important as founders in the planktonic food webs, for upon them most of the animal life of the oceans and shallow seas ultimately depends. By their photosynthetic activity the dinoflagellates and diatoms harness the sun's energy and lock it into organic compounds such as sugars and starch which provide an energy source for the grazers that feed on the phytoplankton.

Zooplankton comprises a wide range of animals. Virtually every known phylum is represented in the sea, and many examples of marine animals have planktonic larvae. Such organisms may be referred to as meroplankton or temporary plankton. Good examples are the developing larvae of bottom-dwellers such as mussels, clams, whelks, polychaete worms, crabs, lobsters and starfish. These larvae ascend into the surface waters and live and feed in a way totally different from that of their adults. Thus the offspring do not compete with the adults for food or living space and the important task of dispersal is achieved by ocean currents. At the end of their plank-tonic lives the temporary plankton have to settle on the seabed and change into adult forms. If the correct substrate is missing then they fail to mature. Often involved physiological and behavioral processes take place before satisfactory settlement can be achieved, and many settling larvae have elaborate mechanisms for detecting textures and chemicals in substrate surfaces.

In addition to the temporary plankton there is a holoplankton: organisms whose entire lives are spent drifting in the sea. Of these the most conspicuous element (around 70 percent) are crustaceans. The most abundant class of planktonic crustaceans are the copepods, efficient grazers of phytoplankton especially in temperate seas. The euphausids make up another very important group of crustacea, and in some regions, eg the southern oceans, they can occur in enormous numbers as "krill," providing the staple diet of the great whales. All these crustaceans have mechanisms for straining the seawater to extract the fine plant cells from it. Other noncrustacean holoplanktonic forms that sieve water for food are the planktonic relatives of the sea squirts. Some rotifers live as herbivores in the surface waters of the sea, but they are a much more important component of the plankton of lakes and rivers. Along with many invertebrate larvae, these herbivorous holoplanktonic forms are important in harvesting the energy contained in the planktonic and pass it on to the carnivorous zooplankton by way of the food webs of the sea's surface.

There are many types of carnivorous zooplankton in the seas of the world and

▲ **Drifters**—centric diatoms protected inside their strong cell walls. In some species these are ornamented or extended to give lift in the water column, functioning like miniature wings.

▼ **Cycle of nutrients and energy** in the sea. This diagram is considerably simplified since many animals obtain food from several levels, forming an intricate "food web."

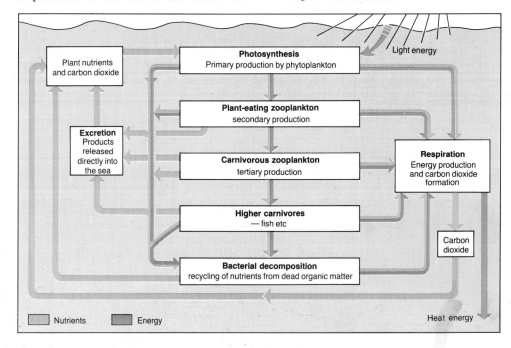

members of many phyla are involved. Protozoans feeding on bacteria or other protozoans occur. Some like the foraminiferans and radiolarians form conspicuous deposits on the seabed after they die because of the enduring nature of their mineralized shells or tests. Cnidarians provide a range of temporary and permanent planktonic carnivores. Many hydroid medusae spend only part of the life cycle of the hydroids in the plankton, while others like the Portuguese man-of-war are permanent plankton dwellers often taking food as large as fishes. The ctenophores, for example *Pleurobrachia* and *Beroë*, are efficient predators of copepods, often outcompeting fishes (for example herrings) for them. Thus they are of economic significance as competitors of commercial fish stocks. Other carnivores include pelagic gastropods, polychaetes and arrow-worms.

The occurrence of certain species in surface waters is taken by oceanographers as an indication of the origins of water currents. Thus in Northwest Europe plankton containing the arrow worm *Sagitta elegans* has been demonstrated to come from the clean open Atlantic whereas water containing *Sagitta setosa* is known to have a coastal origin. Different chaetognaths also appear at different depths in the ocean and are indicative of different animal communities.

AC

▼ **Myriads of animals** under the microscope: a mixture of medusae, copepods, crab larvae, crayfish larvae etc. Some of the larvae are being eaten by the medusae.

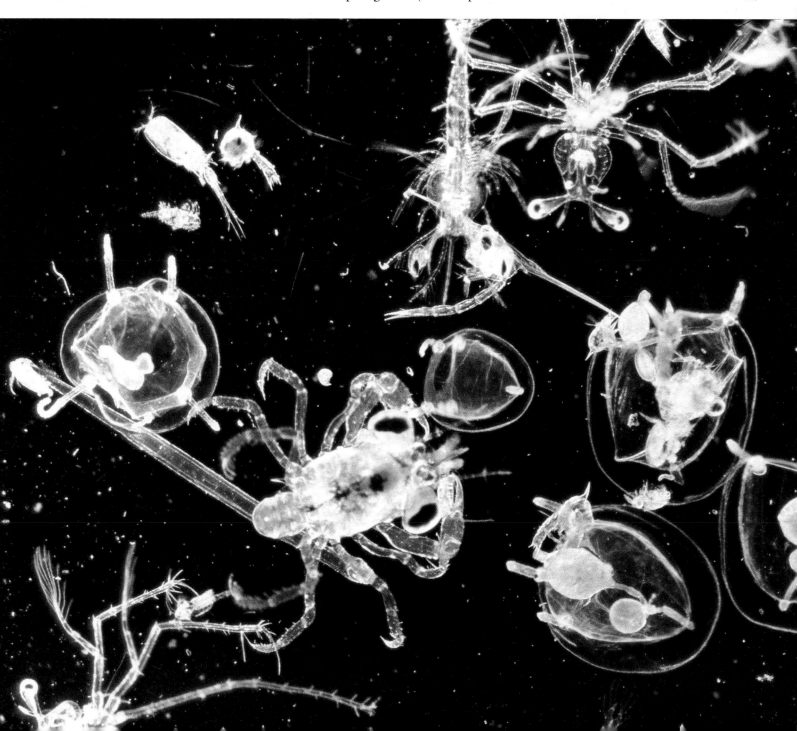

SIMPLE INVERTEBRATES

KINGDOM PROTISTA—SINGLE-CELLED ANIMALS

Subkingdom Protozoa

Flagellates, amoebae, opalinids
Phylum: Sarcomastigophora

Ciliates

Phylum: Ciliophora

Phylum: Apicomplexa—includes malaria, coccidiosis

Phylum: Labyrinthomorpha

Phylum: Microspora

Phylum: Ascetospora

Phylum: Myxospora

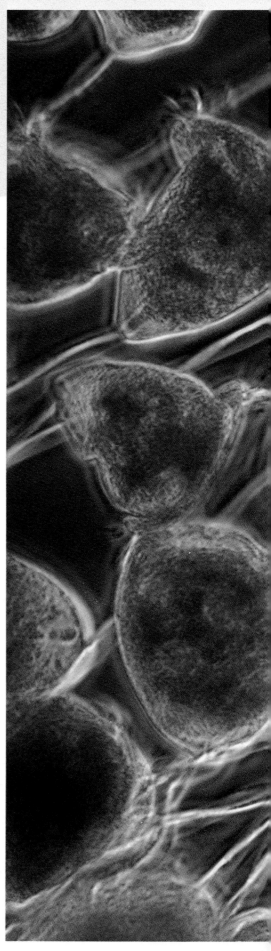

THE expression "Simple Invertebrates" implies an elementary level of organization and superficially this is the case. Here, protozoans, that is single-celled or non-cellular animals, and mesozoans (which have no common name) are simple when compared with other animal architecture. These animals are constructed from one cell or colonies of similar cells, so that there are no tissues and no organs as found in the more highly evolved invertebrates. In virtually every case one cell is capable of fulfilling all the requirements for life including energy acquisition, which may often involve food or prey detection and coordinated movement. This means that the single cell must have a means of detecting stimuli and in turn exciting and controlling a system of movement, for example by the contraction of fibrils or the beating action of cilia and flagella. In addition the cell must be able to control its water content and to have means of resisting drying out. It must be able to reproduce and the offspring must be able to grow in a coordinated fashion and to disperse through the environment. For all these operations to be carried out within one or a few similar cells means that the processes going on inside these cells have to be complex and involved. There is no division of labor in these animals. The result is that protozoan cells tend to be larger and to contain a diversity of parts (nucleus, vacuoles etc) called organelles. All processes necessary for life can be carried out in the one cell.

Because of the high surface area to volume ratio, these animals lose water easily. They are therefore dependent on free water and as such live either in aquatic habitats or in damp places like soil. Many are successful parasites of other animals, flourishing in blood and guts.

In protozoans there are various symmetries, some animals having recognizable front and rear ends, for example *Paramecium*, others being attached by a stalk, for example *Vorticella*, and yet others being totally variable in shape as in amoebae. AC

▶ **Bell-shaped cells,** ringed with thread-like cilia are characteristic of the protozoan ciliate *Vorticella*. These animals are found commonly in freshwater attached by contractile stalks to water plants and larger animals. Minute food particles extracted from the water pass into the vortex created by the beating cilia.

▼ **Are they plants or animals?** This is a question that has often been asked about protozoans, which like true plants contain photosynthetic chloroplasts. Here two dinoflagellates (*Pyrocystis* species) are each in the process of dividing within their original cell walls.

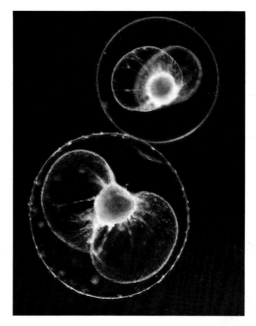

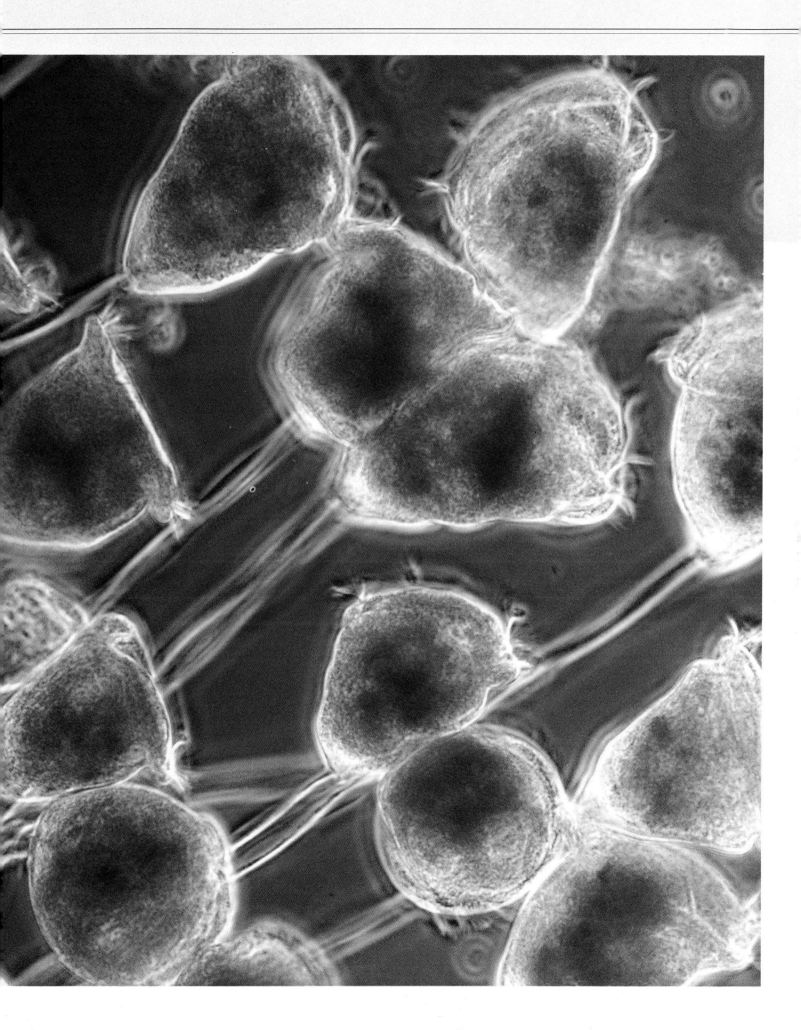

PROTOZOANS

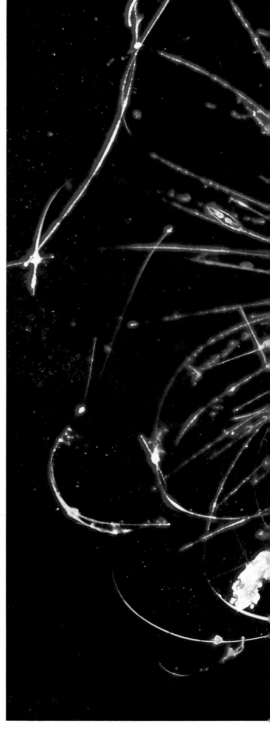

PROTOZOANS live in an unseen world. They are invisible to the naked eye, but occur all around us, beneath us and even within us—they are ubiquitous. Their overriding requirement is for free water so they are mainly found in floating (planktonic) and bottom-dwelling communities of the sea, estuaries and freshwater environments. Some live in the water films around soil particles and in bogs, while others are parasites of other animals, notably causing malaria and sleeping sickness in humans.

Most protozoans are microscopic single-celled organisms living a solitary existence. Some, however, are colonial.

Colony structure varies; in the green flagellate *Volvox* numerous individuals are embedded in a mucilaginous spherical matrix, while in other flagellates, for example *Diplosiga*, a group of individuals occurs at the end of a stalk. Branched stalked colonies are characteristic of some ciliates, such as *Carchesium.*

Many protozoans have evolved a parasitic mode of life involving one or two hosts. Some species live in the guts and urogenital tracts of their hosts. Others have invaded the body fluids and cells of the host, for example the malarial parasite *Plasmodium* which lives for part of its life cycle in the blood and liver cells of mammals, birds and reptiles and the other part in mosquitoes (see pp28–29). Five of the seven protozoan phyla are exclusively parasitic, but among the mainly free-living flagellates, amoebae and ciliates there are some species which have opted for parasitism. Notable among the parasitic flagellates are the hemoflagellates causing various forms of trypanosomiasis, including sleeping sickness. The opalinid gut parasites of frogs and toads are another example. In most multicellular animals the essential life processes are carried out by specialized tissues and organs. In protozoans all life processes occur in the single cell and the building blocks for these specialized functions are small tubular structures termed microtubules. These have reached their most complex organization in the ciliates, the most advanced of all protozoans. Apart from a few flagellate groups, the ciliates are the only protozoans to possess a true cell mouth or cytostome. In addition they typically have two types of nuclei (dimorphism), each performing a different role. The macronucleus, which is often large and may be round, horseshoe shaped, elongated or resemble a string of beads, controls normal physiological functioning in the cell, while the micronucleus is concerned with the replication of genetic material during reproduction. It is quite common for a ciliate to possess several micronuclei. Other protozoans have nuclei of one type only, although some species may have several. The exceptions are the foraminiferans which show nuclear dimorphism at some stages in their life cycle. The most widely known and researched ciliates are species of the genera *Paramecium* and *Tetrahymena*, but these represent only a minute fraction of the 7,000 species so far described by scientists.

The majority of protozoans feed on bacteria, algae, other protozoans, microscopic animals and in the case of parasites on host tissue, fluids and gut contents. Their diet incorporates complex organic compounds of nitrogen and hydrogen, and they are said to be heterotrophic. Some flagellates, such

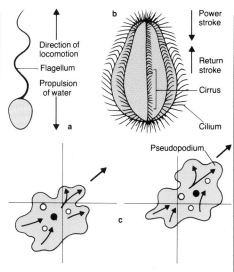

Direction of
locomotion

Flagellum

Propulsion
of water

a

b

Power
stroke

Return
stroke

Cirrus

Cilium

Pseudopodium

c

▲ **Movement in protozoans** is by three means.
(1) By a single flagellum which, like a propeller,
pulls the cell through the water. (2) By rows of
cilia which are coordinated to act like the oars
of a rowboat, having a power stroke and a
recovery stroke. (3) By amoeboid movement
whereby pseudopodia ("false feet") are
extended and the rest of the body flows into
them. Pseudopodia are also involved in prey
capture.

◄ **Like exploding fireworks,** the slender
filaments of a microscopic foraminiferan
(*Globigerinoides* species) catch the light under
the microscope. The filaments (pseudopodia)
radiate out from the rest of the cell inside a hard
shell or test. They are used to trap food and for
movement.

▼ **Protozoan forms.** (1) *Actinophrys*
(heliozoan). (2) *Opalina* (opalinid). (3) *Acineta*
(suctorian). (4) *Elphidium* (foraminiferan).
(5) *Euglena* (phytoflagellate). (6) *Trypanosoma*
(hemoflagellate). (7) *Hexacontium* (radiolarian).

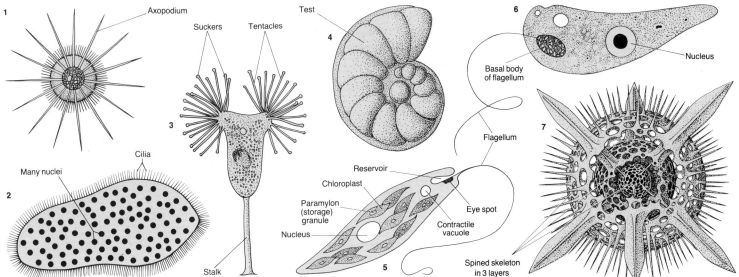

1

Axopodium

Suckers   Tentacles

Test

4

6

Nucleus

Basal body
of flagellum

Flagellum

3

Cilia

Many nuclei

2

Stalk

Reservoir

Chloroplast

Paramylon
(storage)
granule

Nucleus

Eye spot

Contractile
vacuole

5

7

Spined skeleton
in 3 layers

as *Euglena* and *Volvox*, however, possess photosynthetic pigments in chloroplasts. These protozoans are capable of harnessing the sun's radiant energy in the chemical process of photosynthesis to construct complex organic compounds from simple molecules—they are said to be autotrophic. A number of such flagellates must, however, combine autotrophy with heterotrophy, in varying degrees. Such organisms sit on the boundary between an animal and plant-like nutrition (see p12).

Many protozoans are simply bound by the cell wall, but skeletal structures in the form of secreted shells or tests are common among the amoebae and usually have a single chamber. The exclusively marine foraminiferans, however, are exceptional in having shells with numerous chambers. Shells and tests may be formed of calcium carbonate or silica, or from organic substances such as cellulose or chitin.

Most free-living and some parasitic species need to move around their environment to feed, to move toward and away from favorable and unfavorable conditions, and in some cases special movement is required in reproductive processes. The various protozoan groups achieve movement using different structures.

Members of the subphylum Sarcodina (including the amoebae) produce so-called pseudopodia—flowing extensions of the cell. These may be extended only one at a time, as in *Naegleria* or several at a time, as in *Amoeba proteus*, *Arcella* and *Difflugia*. Heliozoan sarcodines, which resemble a stylized sun, possess long slender pseudopodia, called axopodia, which radiate from a central cell mass. Each axopodium is supported by a large number of microtubules arranged in a parallel fashion along the longitudinal axis. Heliozoans move slowly, rolling along by repeatedly shortening and lengthening the axopodia. A well-known example of these so-called sun organisms is *Actinosphaerium*. The foraminiferans, for example *Elphidium*, which bear complex chambered shells, have a complicated network of pseudopodial strands which branch and fuse with each other to produce a linking complex of what are termed reticulopodia. Like the axopodia of heliozoans the reticulopodia of foraminiferans are supported by microtubules.

The other means of movement is by the beating action of the filamentous cilia and flagella, which are permanent outgrowths of the cell rather than, like pseudopodia, its temporary pseudopodial extensions.

Cilia and flagella are structurally similar,

but cilia are shorter. Normally flagellates carry only one or two flagella, while in the ciliates the cilia are numerous and usually arranged in ordered rows each called kinety. The number of kinety is constant in each species and is used as an aid in identification. In some cases cilia may fuse to form cirri, which resemble short thick hairs, or structures which are sail-like. Each cilium and flagellum is about $0.15–0.3\mu$m in diameter and is supported by a core (axoneme) made up of two centrally positioned microtubules surrounded and joined by cross-bridges to nine double microtubules. This $9+2$ arrangement of microtubules is common in cilia and flagella throughout the living world—from amoebae to invaders of lung linings of humans. Movement in cilia and flagella involves the passage of waves along

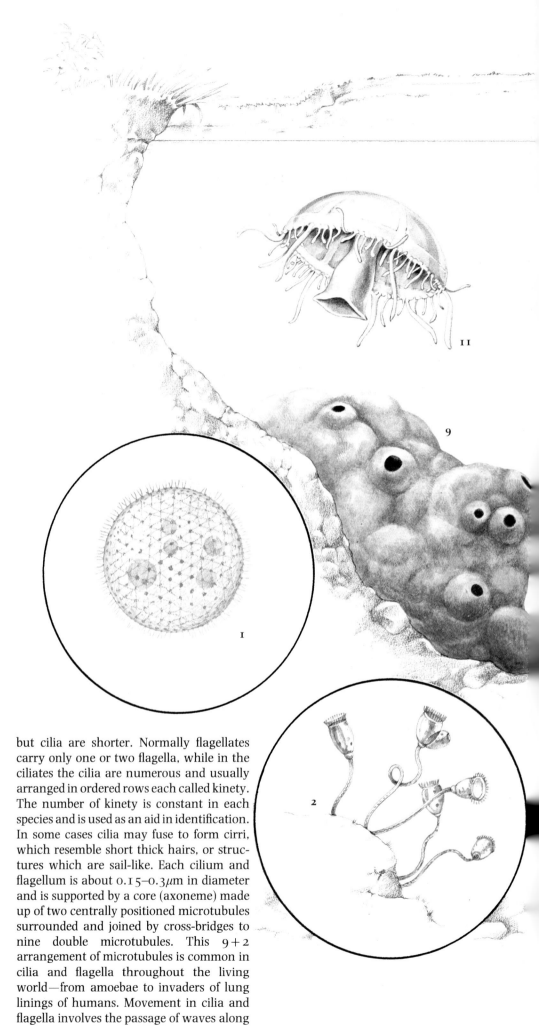

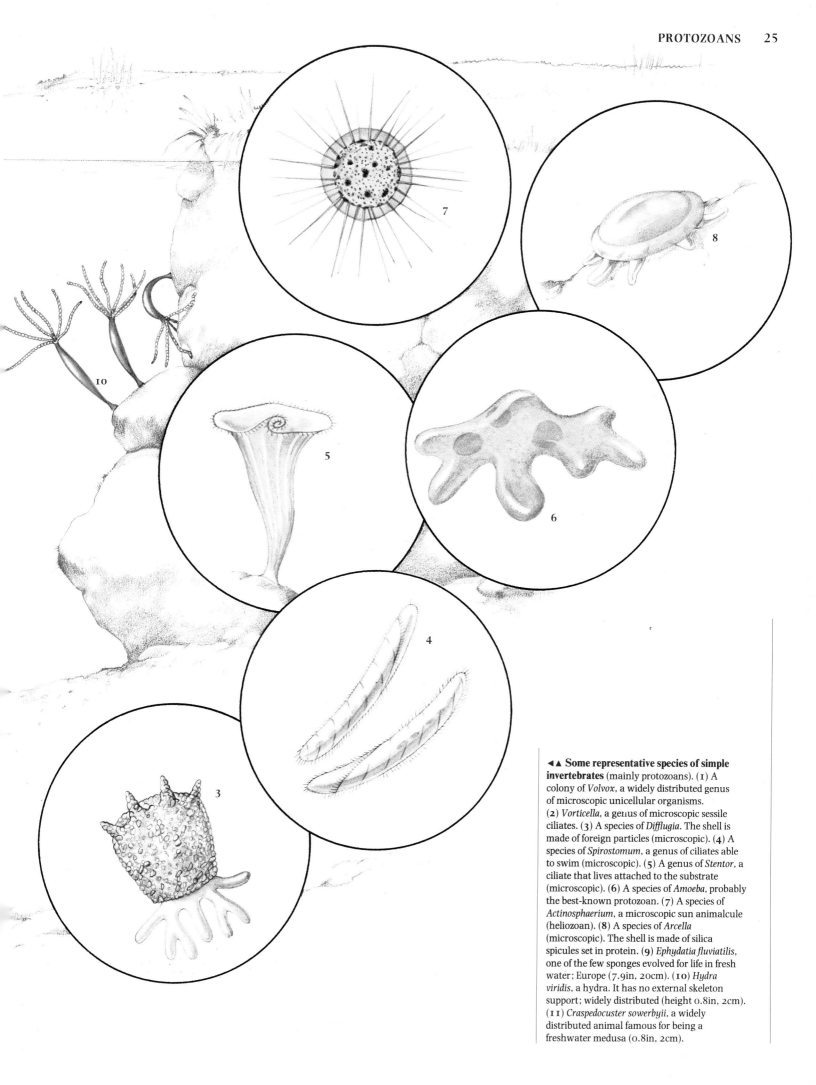

◄▲ **Some representative species of simple invertebrates** (mainly protozoans). (1) A colony of *Volvox*, a widely distributed genus of microscopic unicellular organisms. (2) *Vorticella*, a genus of microscopic sessile ciliates. (3) A species of *Difflugia*. The shell is made of foreign particles (microscopic). (4) A species of *Spirostomum*, a genus of ciliates able to swim (microscopic). (5) A genus of *Stentor*, a ciliate that lives attached to the substrate (microscopic). (6) A species of *Amoeba*, probably the best-known protozoan. (7) A species of *Actinosphaerium*, a microscopic sun animalcule (heliozoan). (8) A species of *Arcella* (microscopic). The shell is made of silica spicules set in protein. (9) *Ephydatia fluviatilis*, one of the few sponges evolved for life in fresh water; Europe (7.9in, 20cm). (10) *Hydra viridis*, a hydra. It has no external skeleton support; widely distributed (height 0.8in, 2cm). (11) *Craspedocuster sowerbyii*, a widely distributed animal famous for being a freshwater medusa (0.8in, 2cm).

them from one axis to the other. Most flagella move in two-dimensional waves, while cilia move in three-dimensional patterns coordinated into waves which result from fluid forces (hydrodynamic forces) acting on the automatic beating of each cilium.

Reproduction in the protozoans does not usually involve sex or sexual organelles—it is asexual. In most free-living species asexual reproduction occurs by a process called binary fission, whereby each reproductive effort results in two identical daughter cells by the division of a parent cell. In the flagellates, including the parasitic species, the plane of division is longitudinal, while in the ciliates it is normally transverse and prior to division of the cytoplasm the mouth is replicated. The amoebae do not normally have a fixed plane for division. In shelled and testate species the process is complicated by the need to replicate skeletal structures. In testate species of amoeba such as *Difflugia*, cytoplasm destined to become the daughter is extruded from the aperture of the parent test. Preformed scales in the cytoplasm then

Prey Capture in Carnivorous Protozoans

Carnivorous protozoans prey on other protozoans, rotifers, members of the Gastrotricha and small crustaceans. The mode of capture and ingestion is often spectacular and frequently the prey are larger than the predators.

Among the ciliates the sedentary Suctoria have lost their cilia, which have been replaced by tentacles, each of which functions as a mouth. When other ciliates, such as *Colpidium*, collide with a tentacle, they stick to it. Other tentacles move toward the prey and also attach. The cell wall of the prey is perforated at the sites of attachment and the prey cell contents are moved up the tentacle by microtubular elements within the tentacle. A single suctorian, for example *Podophrya*, can feed simultaneously on four or five prey. *Didinium nasutum* is a ciliate that feeds exclusively on *Paramecium*, which it apprehends using extrudable structures called pexicysts and toxicysts. The former hold the prey while the latter penetrate deeply into it releasing poisons. *Didinium* consumes the immobilized prey whole, its body becoming distended by the ingested *Paramecium*.

The heliozoan *Actinophrys* also feeds on ciliates which are captured, on contact, by the radiating axopodia. Once attached, the prey is progressively engulfed by a large funnel-shaped pseudopodium produced by the cell body. Occasionally when an individual *Actinophrys* has captured a large prey, other *Actinophrys* may fuse with the feeding individual to share the meal. In such instances, after digesting their ciliate victim, the heliozoan predators separate again.

The foraminiferan *Pilulina* has evolved into a living pit-fall trap. This bottom-dwelling species builds a bowl-shaped shell or test with mud, camouflaging the pseudopodia across the entrance. When copepod crustaceans blunder onto the pseudopodia they get stuck and are drawn down into the animal. The radiolarians, which possess a silica-rich internal skeleton, deal with copepod prey by extending the wave-flow along the axopodia to the broad surfaces of the prey's exoskeleton, and rupturing the prey by force. The axopodia then penetrate and prize off pieces of flesh which are directed down the axopodia to the main cell body for ingestion.

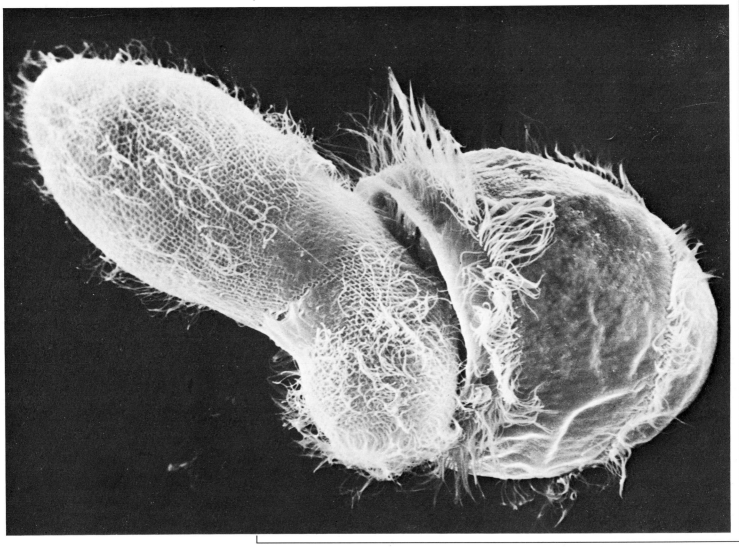

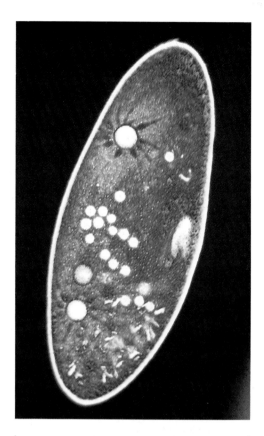

▲ **Slipper animalcules** (*Paramecium* species) mostly live in fresh and stagnant water, feeding on bacteria and particles of plant food. Here food in the oral groove shows stained orange and the star-shaped objects at top and bottom are contractile vacuoles which collect excess water and expel it from the body.

◄ **A struggle for life and death,** a *Paramecium* about to be devoured by a *Didinium*.

▼ **Divide and reproduce.** Asexual reproduction in protozoans is by simple division (binary fission). In flagellates (1) (eg *Euglena*) it is longitudinal. In ciliates (2) (eg *Tetrahymena*) division is complicated by the replication of the oral apparatus. In shelled amoebae (3) (eg *Euglypha*) the daughter is extruded from the parent. Amoebae (4) have no fixed plane of division.

form a test around the extruded cytoplasm. When the process is complete the two amoebae separate.

Most free-living species normally reproduce asexually provided conditions are favorable. Sexual reproduction is usually only resorted to in adversity, such as drying up of the aquatic medium when the normal cells would not survive. The ability to undergo a sexual phase is not widespread in the amoebae and flagellates and is restricted to a limited number of groups. Some species may never have reproduced sexually in their evolutionary history, but others may have lost sexual competence. Both isogamous (reproductive cells or gametes alike) and the more advanced anisogamous (reproductive cells or gametes dissimilar) forms of sexual reproduction occur.

The foraminiferans are unusual among free-living species in having alternation of asexual and sexual generations. Here each organism reproduces asexually to produce many amoeba-like organisms which secrete shells around themselves. When mature these produce many identical gametes which are usually liberated into the sea, where they fuse in pairs to produce individuals which in turn secrete a shell, grow to maturity and repeat the cycle.

Almost all of the ciliates are capable of sexual reproduction by a process called conjugation, which does not result in an immediate increase in numbers. The function of conjugation is to facilitate an exchange of genetic materials between individuals. During this process two ciliates come together side by side and are joined by a bridge of cell contents (cytoplasm). A complex series of divisions of the micronucleus occurs, including a halving of the pairs of chromosomes (or meiosis).

In the final stages a micronucleus passes from each individual into the other. Essentially the micronuclei are gametes. Each received micronucleus fuses with an existing micronucleus in the recipient. The ciliates separate and after further nuclear divisions eventually undergo binary fission.

All members of the parasitic phyla, except for some groups in the Apicomplexa, produce spores at some stages in their life cycles. The Apicomplexa contains a number of parasites of medical and veterinary importance, including the malarial parasites *Plasmodium* and the *Coccidia* responsible for coccidiosis in poultry. Some species, like *Plasmodium*, have complex life cycles involving two hosts with an alternation of sexual and asexual phases. In *Plasmodium* the sexual phase is initiated in humans and is completed in the mosquito; following this many thousands of motile spores (sporozoites) are reproduced which are infective to humans and are transmitted when the mosquito feeds. In humans repeated phases of multiple asexual division take place in the red blood cells and liver cells (see p28). The phyla Microspora, Ascetospora and Myxospora are parasites of a wide range of vertebrates and invertebrates, while the Labyrinthomorpha parasitize algae.

The ecology of protozoans is very complex, as one would expect in a group of ubiquitous organisms. They are found in the waters and soils of the world's polar regions. Some have adapted to warm springs and there are records of protozoans living in waters as warm as 154°F (68°C). Protozoans occur commonly in planktonic communities in marine, brackish and freshwater habitats, and also in the complex bottom-dwelling (benthic) communities of these environments. Little is known about protozoans in the marine deeps, but there is a record of foraminiferans living at 13,000ft (4,000m) in the Atlantic. Ciliates, flagellates and various types of amoebae are also common in soils and boggy habitats.

Since many protozoans exploit bacteria as a food source they form part of the decomposer food web in nature. Recent research suggests that protozoans may stimulate the rate of decomposition by bacteria and thus enhance the recycling of minerals such as phosphorus and nitrogen. The exact mechanism is not entirely clear, but protozoans grazing on bacteria may maintain the bacterial community in a state of physiological youth and hence at the optimum level of efficiency. There is also evidence that some protozoans secrete a substance that promotes the growth of bacteria.

JL-P

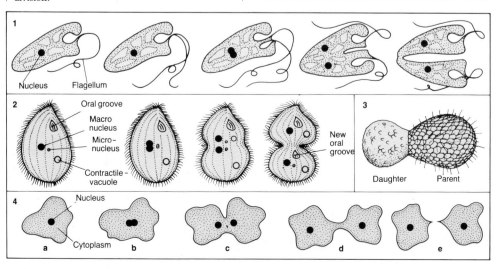

Malaria and Sleeping Sickness

Protozoan diseases of man

One million people die each year in Africa from malaria—this statistic exemplifies the virulence of protozoan diseases. The most serious pathogens are *Plasmodium*, which produces malaria, and various trypanosome species responsible for diseases broadly called trypanosomiasis or sleeping sickness.

Malaria is caused by four species of the genus *Plasmodium*. The life cycle is similar in each species but there are differences in disease pathology. *Plasmodium falciparum* causes malignant tertian malaria and accounts for about 50 per cent of all malarial cases. It attacks all red blood cells (erythrocytes) indiscriminately so that as many as 25 per cent of the erythrocytes may be infected. In this species stages not involving the erythrocytes do not persist in the liver, so that relapses do not occur. *Plasmodium vivax* produces benign tertian malaria, which invades only immature erythrocytes so that the level of cells infected is low. Here, however, other stages remain in the liver, causing relapses. Benign tertian malaria is responsible for approximately 45 per cent of malarial infections. The other two species are relatively rare. *Plasmodium malariae*, causing quartan malaria, attacks mature red blood cells and has persistent stages outside the blood cells. Little is known about *P. ovale* because of its rarity.

The diseases are named after the fevers which the parasites cause, tertian fevers occurring every three days or 48 hours and quartan fevers every four days or 72 hours. The naming practice is based on the Roman system of calling the first day one, whereas we would call the first day nought.

Once inside an erythrocyte, the parasite feeds on the red blood cell contents and grows. When mature it undergoes multiple asexual fission to produce many individuals called merozoites which by an unknown mechanism rupture the erythrocyte and escape into the blood plasma. Each released merozoite then infects another erythrocyte. The asexual division cycle in the red blood cells is well synchronized so that many erythrocytes rupture together—a phenomenon responsible for the characteristic fever which accompanies malaria. The exact mechanism producing the fever is not fully understood, but it is believed to be caused by substances (or a substance), possibly derived from the parasite, which induce the release of a fever-producing agent from the white blood cells, which fight the disease. When the parasite has undergone a series of asexual erythrocytic cycles, some individuals produce the male and female gametocytes which are the stages infective to the mosquito host. The stimulus for gametocyte production is unknown.

Malaria is still one of the greatest causes of death in humans. Tens of millions of cases are reported each year and many are fatal. Successful control measures are available, and in countries such as the USA, Israel and Cyprus the disease has been eradicated. In Third World countries, however, control measures have little impact on malaria. Broadly, eradication programs involve the use of drugs to treat the disease in humans, and a series of measures aimed at breaking the parasite's life cycle by destroying the intermediate mosquito host.

Like many insects, the mosquito has an aquatic larval stage. The draining of swamps and lakes deprives the mosquito of an environment for breeding and its larval development, but residual populations continue to breed in irrigation canals, ditches and paddy fields. Spraying of oil on the water surface asphixiates the larvae, which have to come to the water surface periodically to breathe. The poison Paris Green can effectively kill larvae when added to the water. Biological control using fish predators of mosquito larvae, such as the guppy, aid in reducing larval populations.

Adult mosquitoes can be killed by spraying houses with various insecticides such as hexachlorocyclohexane and dieldrin. In the past DDT was very successful but its toxicity

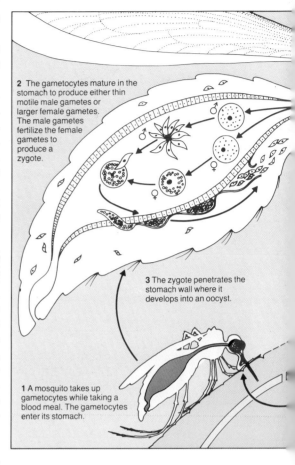

2 The gametocytes mature in the stomach to produce either thin motile male gametes or larger female gametes. The male gametes fertilize the female gametes to produce a zygote.

3 The zygote penetrates the stomach wall where it develops into an oocyst.

1 A mosquito takes up gametocytes while taking a blood meal. The gametocytes enter its stomach.

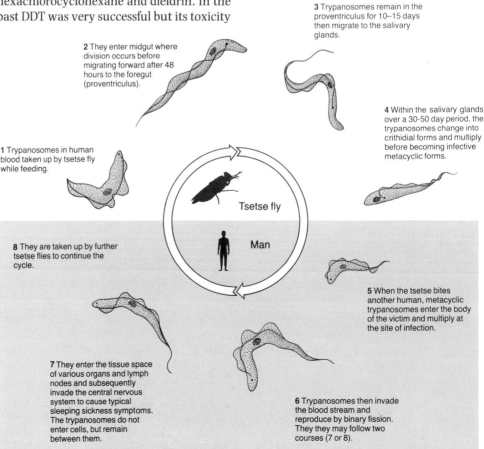

1 Trypanosomes in human blood taken up by tsetse fly while feeding.

2 They enter midgut where division occurs before migrating forward after 48 hours to the foregut (proventriculus).

3 Trypanosomes remain in the proventriculus for 10–15 days then migrate to the salivary glands.

4 Within the salivary glands over a 30-50 day period, the trypanosomes change into crithidial forms and multiply before becoming infective metacyclic forms.

5 When the tsetse bites another human, metacyclic trypanosomes enter the body of the victim and multiply at the site of infection.

6 Trypanosomes then invade the blood stream and reproduce by binary fission. They they may follow two courses (7 or 8).

7 They enter the tissue space of various organs and lymph nodes and subsequently invade the central nervous system to cause typical sleeping sickness symptoms. The trypanosomes do not enter cells, but remain between them.

8 They are taken up by further tsetse flies to continue the cycle.

Tsetse fly

Man

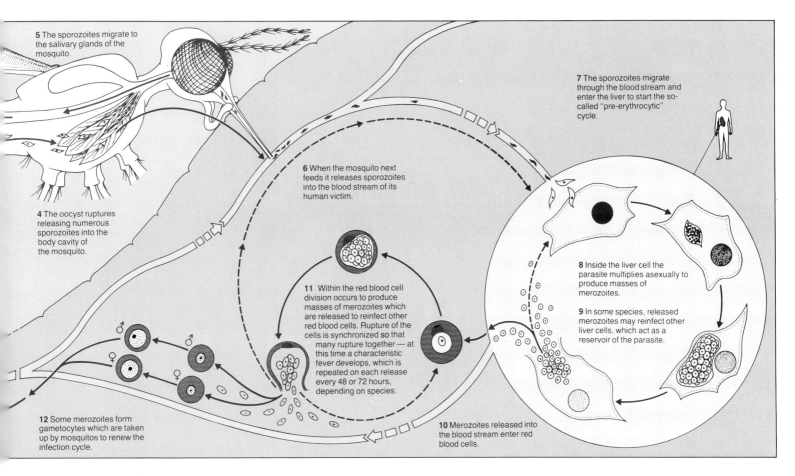

5 The sporozoites migrate to the salivary glands of the mosquito.

7 The sporozoites migrate through the blood stream and enter the liver to start the so-called "pre-erythrocytic" cycle.

6 When the mosquito next feeds it releases sporozoites into the blood stream of its human victim.

4 The oocyst ruptures releasing numerous sporozoites into the body cavity of the mosquito.

8 Inside the liver cell the parasite multiplies asexually to produce masses of merozoites.

9 In some species, released merozoites may reinfect other liver cells, which act as a reservoir of the parasite.

11 Within the red blood cell division occurs to produce masses of merozoites which are released to reinfect other red blood cells. Rupture of the cells is synchronized so that many rupture together — at this time a characteristic fever develops, which is repeated on each release every 48 or 72 hours, depending on species.

12 Some merozoites form gametocytes which are taken up by mosquitos to renew the infection cycle.

10 Merozoites released into the blood stream enter red blood cells.

▲ ◄ **Life cycles of malaria** ABOVE **and sleeping sickness** LEFT. A fifth of the world's four-billion population is probably threatened by malaria, while in Africa 50 million people are exposed to sleeping sickness or trypanosomiasis.

▼ **Hidden killer.** A mass of the pre-erythrocytic stage of *Plasmodium* (malaria) developing in the human liver.

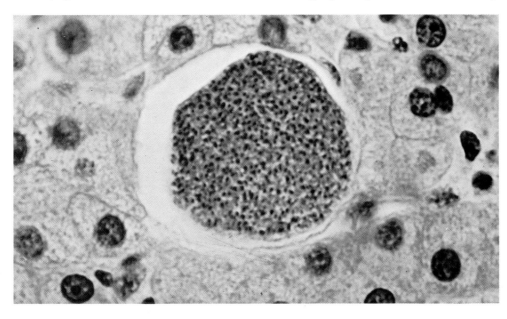

to higher animals now precludes its use. Biological control measures involve releasing sterile male mosquitoes into the population, thereby decreasing reproduction rates, and the introduction of bacterial, fungal and protozoan pathogens of the mosquito.

Chemotherapy in humans involves four broad categories of treatment. Firstly, there are prophylactic drugs, such as Proguanil, which taken on a regular basis prevent recurring erythrocytic infections. Secondly, there are drugs such as Chloroquine which destroy the blood stages of the parasite. Thirdly there are drugs which destroy gametocytes. Lastly there are drugs which, when taken up by the mosquito during feeding on humans, prevent further development of the parasite in the insect. Drug resistance by *Plasmodium* does occur; *P. falciparum* has become resistant to Chloroquine in some parts of Africa and South America, and has to be treated by a combination of quinine and sulfonamides.

The flagellates *Trypanosoma rhodesiense* and *T. gambiense* cause African trypanosomiasis or sleeping sickness. The two-host life cycle involves a tsetse fly (genus *Glossina*) and humans. In man, the trypanosomes live in the blood plasma and lymph glands, progressing later to the cerebrospinal fluid and brain. The disease is typified by mental and physical apathy and a desire to sleep. The disease is fatal if untreated, *T. rhodesiense* running a more acute course than *T. gambiense*. Control measures include insecticide use, introduction of sterile males, clearing vegetation in which *Glossina* spends the whole of its life cycle, and the use of drugs in humans. Control is complicated by the fact that *T. rhodesiense* also infects game animals, so that a reservoir population of the parasite persists.   JL-P

Subkingdom Parazoa
Sponges
Phylum: Porifera

Subkingdom Mesozoa
Mesozoans
Phylum: Mesozoa

Subkingdom Metazoa
Sea anemones, jellyfishes and their allies
Phylum: Cnidaria

Comb jellies
Phylum: Ctenophora

Endoprocts
Phylum: Endoprocta (Kamptozoa)

Rotifers
Phylum: Rotifera

Kinorhynchs
Phylum: Kinorhyncha

Gastrotrichs
Phylum: Gastrotricha

Lampshells
Phylum: Brachiopoda

Moss animals or sea mats
Phylum: Bryozoa (Ectoprocta)

Horseshoe worms
Phylum: Phoronida

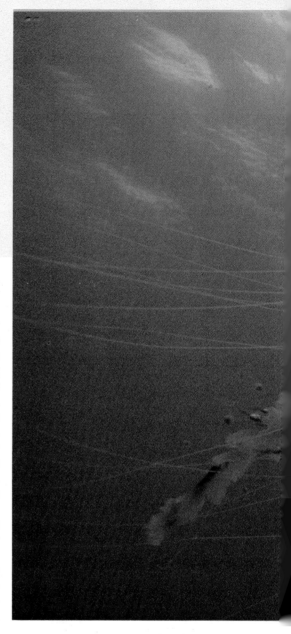

LIFE appears to have originated in the oceans and virtually every group of animals known to man has representatives living in the sea. A few of these groups such as the sponges (phylum Porifera), hydroids (phylum Cnidaria) and the moss animals or sea mats (phylum Bryozoa) have invaded fresh water, but none has been very successful there.

Life in water offers all sorts of possibilities to animals. The drifting communities of plankton teeming in the surface waters of the open seas offer vast resources of food to those swimming animals that are able to utilize microscopic suspended food matter. The sea bed (and to a lesser extent the lake and river floor) provides a variety of habitats—hard stones, rocks, soft sands and muds—which can give support to animals if they live attached to the substrate.

Solitary invertebrates include many that need a firm base for attachment, for example sponges, hydroids and sea anemones, corals, endoprocts and sea mats. Invariably these animals adopt a plant-like growth form which confused the early naturalists. Many of them are colonial too; that is, a number of individuals live inside a common shared body mass. Being attached to a fixed object, rock, pebble or man-made structure has many problems, and these organisms frequently display either an asymmetry (as in the sponges) or radial symmetry (as in the hydroids, comb jellies, endoprocts etc). The organisms cannot move, either to gain food or to escape predation. This means that they have to exploit naturally occurring currents to bring them suspended particles of food. (Some other invertebrate groups like the bivalve mollusks (mussels, clams etc) can generate suitable currents by special pumping systems themselves.) Many sedentary organisms then resort to crowns of tentacles which will act as filters for food-gathering

and function simultaneously as gills in respiration. Radial symmetry is ideally suited to this type of function.

Defense against predation is usually achieved either by inedibility (for example sponges with a bad taste and hard spicules) or by specific weaponry (such as the stinging cells of hydroids and jellyfishes). In others the body is housed in a box-like external skeleton into which the delicate parts of the crown can be withdrawn (for example sea mats). The tendency of such organisms to form colonies is an asset since it allows some individuals to become specialized at certain roles, such as food-gathering, while others pursue functions in reproduction or defense.

Attached animals are often referred to as sessile, but sessile strictly means attached by a stalk, not directly encrusting the substrate. Not all sedentary animals are of course attached immovably by a holdfast (a grasping structure which resembles in appearance, but not in function, the roots of a plant). Sea anemones grip by means of a sole-like sucker and can creep around over rocks. Some can burrow in sand.

The free-swimming animals present a direct contrast, with their often elaborate systems of movement. These can range from pulsing muscular systems (jellyfishes), single-cilia systems (rotifers and gastrotrichs) and compound ciliary systems (comb jellies). These examples are drawn from those free-swimming animals that are closely or more distantly related to the sedentary bottom-living types also dealt with here. In the higher groups of invertebrates more elaborate systems will be shown. The links between the sedentary and free-swimming groups here treated are evidenced by the free-swimming medusae of the stalked hydroids which essentially act as reproductive and dispersal phases in the hydroids' life cycles.

▲ **Deadly beauty.** Contact with the trailing tentacles of this Compass jellyfish (*Chrysaora hysoscella*) means certain death to small swimming organisms which are stunned by its stinging cells.

▶ **A sit-and-wait predator.** Sea anemones rest attached to the substrate waiting for prey to blunder into their outstretched tentacles. They are often well camouflaged against detection by eye.

IMMING INVERTEBRATES

The comb jellies are the most remarkable form of swimming animals, occupying an interesting evolutionary position somewhat between the radial symmetry of the hydroids etc and the true bilateral body form of the flatworms. Comb jellies are virtually all true plankton-dwellers whose almost invisible transparent bodies drift in the oceans trailing tentacles like fishing lines. These are armed with unique lasso cells that explode and ensnare their microscopic prey. The small and relatively insignificant rotifers, gastrotrichs and kinorhynchs have some importance in the food web but are otherwise of academic interest only. Many rotifers are sedentary, living in tubes or attached to other organisms like plants. In a few planktonic types their "wheel organs" serve to propel them through the water.

AC

SPONGES

Phylum: Porifera
Sole phylum of subkingdom Parazoa.
About 5,000 species in 790 genera and 80 families.
Distribution: worldwide, marine and freshwater, intertidal to deep sea.
Fossil record: originated in Cambrian 570–500 million years ago; 390 genera identified from Cretaceous (135–65 million years ago).
Size: from microscopic to 6.6ft (2m); the largest sponges occur in the Antarctic and the Caribbean.
Features: form variable; solitary or colonial; mostly porous, filter-feeding organisms mostly attached direct to substrate, without "stem"; lack organs and have little in way of definite tissues, but with complex array of cell types; skeleton lacking or of siliceous or calcareous spicules, or of organic spongin fibers; generally hermaphrodite; sexual and asexual reproduction.

Glass or siliceous sponges

Class Hexactinellida (Hyalospongiae)
About 600 species. Marine, below tidal levels but more common in deeper waters. Skeleton of complex silica spicules, with basic pattern of 6 rays. Includes *Aphrocallistes*, *Euplectella aspergillum* (**Venus' flower basket**), *Holascus*, *Pheronema*.

Calcareous sponges

Class Calcarea
About 400 species. Marine. Skeleton of calcareous spicules which are needle-like or 3- or 4-rayed. Includes *Acyssa*, *Clathrina*, *Leucilla*, *Leucosolenia*, *Scypha*.

Typical sponges

Class Demospongiae
About 4,000 species. Marine and freshwater. Skeleton lacking or of silica spicules, spongin fibers or both. Includes: *Aplysina*, *Cliona*, *Cribochalina vasculum* (**Caribbean sponge**), *Ephydatia*, *Haliclona*, *Hippospongia communis* (**bath sponge**), *Neofibularia nolitangere* (**Caribbean fire sponge**), *Siphonodictyon*, *Spongia officinalis* (**bath sponge**), *Spongilla*.

Coralline sponges

Class Sclerospongiae
About 15 species. Marine, in tropical, shallow, subtidal caves or underneath corals. Skeleton with calcareous base and entrapped silica spicules and organic fibers. Sponge forms thin layer over calcareous base. Includes: *Ceratoporella*, *Stromatospongia*.

▶ **Like a cluster of smokestacks,** a purple column or tube sponge (*Verengia lacunosa*) rises from the seabed. The large exhalent opening (osculum) can be seen at the top of the lower column.

THE humble bath sponge has been used by people since earliest times, particularly in the Mediterranean region. Bath sponge species are the best known of a group of animals whose relationship to other organisms is a matter of debate. Until the early 19th century sponges were regarded as plants, but they are now generally considered to be a group (phylum Porifera) of animals placed within their own subkingdom, the Parazoa. They probably originated either from flagellate protozoans or from related primitive metazoans.

Sponges range in size from the microscopic up to 6.6ft (2m). They often form a thin incrustation on hard substrates to which they are attached, but others are massive, tubular, branching, amorphous or urn-, cup- or fan-shaped. They may be drab or brightly colored, the colors derived from mostly yellow to red carotenoid pigments.

All sponges are similar in structure. They have a simple body wall containing surface (epithelial) and linking (connective) tissues, and an array of cell types, including cells (amoebocytes) that move by means of the flow of protoplasm (amoeboid locomotion). Amoebocytes wander through the inner tissues, for example, secreting and enlarging the skeletal spicules and laying down spongin threads. Sponges are not totally immovable but the main body may show very limited movement through the action of cells called myocytes, but they often remain anchored to the same spot.

Although sponges are soft-bodied, many are firm to touch. This solidity is due to the internal skeleton comprised of hard rod- or star-shaped calcareous or siliceous spicules and/or of a meshwork of protein fibers called spongin, as in the bath sponge. Spicules may penetrate the sponge surface of some species and cause skin irritation when handled.

Sponges are filter-feeders straining off bacteria and fine detritus from the water. Oxygen and dissolved organic matter are also absorbed and waste materials carried away. Water enters canals in the sponge through minute pores in their surface and moves to chambers lined by flagellate cells called choanocytes or collar cells. The choanocytes ingest food particles, which are passed to the amoebocytes for passage to other cells. Eventually the water is expelled from the sponge surface, often through volcano-like oscules at the surface. Water is driven through the sponge mainly by the waving action of flagella borne by the choanocytes.

Sponges reproduce asexually by budding off new individuals, by fragmentation of

parts which grow into new sponges and, particularly in the case of freshwater sponges, by the production of special gemmules. These gemmules remain within the body of the sponge until it disintegrates, when they are released. In freshwater sponges, which die back in winter in colder latitudes, the gemmules are very resistant to adverse conditions, such as extreme cold. Indeed, they will not hatch unless they have undergone a period of cold.

In sexual reproduction, eggs originate from amoebocytes and sperms from amoebocytes or transformed choanocytes, usually at different times within the same individual. The sperms are shed into the water, the eggs often being retained within the parent, where they are fertilized. Either solid (parenchymula) or hollow (amphiblastula) larvae may be produced; many swim for up to several days, settle, and metamorphose into individuals or colonies that feed and grow. Others creep on the substrate before metamorphosis. Some mature Antarctic sponges have not grown over a period of 10 years.

▲ **A free run.** This encrusting red sponge spreads over the rock face particularly well in low light intensity because competition with algae for space is reduced when the illumination is too poor for the plants to flourish.

▼ **The three basic body forms of sponges.** (1) (2) (3) Sponges are filter feeders straining off bacteria and fine detritus from the water. Oxygen and dissolved organic matter are also absorbed and waste materials carried away. Water enters canals in the sponge through minute pores in their surface and moves to chambers lined by flagellate cells called choanocytes or collar cells (4). The choanocytes ingest food particles, which are passed to the amoebocytes for passage to other cells. Eventually the water is expelled from the sponge surface, often though volcano-like oscules at the surface. Water is driven through the sponge mainly by the beating action of the flagellae borne on the choanocytes. Choanocytes may line the body cavity (1), or the wall is folded so these cells line pouches (1, 2) connected to more complicated canal systems.

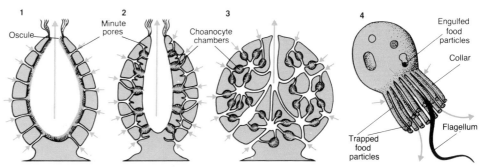

1   Oscule

2   Minute pores

3   Choanocyte chambers

4   Engulfed food particles   Collar   Trapped food particles   Flagellum

Sponges are found in large numbers in all the seas of the world. They occur in greatest abundance on firm substrates, relatively few being adapted to life on unstable sand or mud. Their vertical range includes the lowest part of the shore subject to tidal effects and extends downwards as far as the abyssal depths of 27,000ft (8,600m). One family of siliceous sponges, the Spongillidae, has invaded freshwater lakes and rivers throughout the world.

Sponges living between tide marks are typically confined to parts of the shore that are seldom exposed to the air for more than a very short period. Some occur a little higher up the shore, but these are only found in shaded situations or on rocks facing away from the sun.

Some sponges are killed by even a relatively short exposure to air, and it is in the shallow waters of the continental shelf that sponges achieve their greatest abundance in terms both of species and individuals.

Cavernous sponges are frequently inhabited by smaller animals, some of which cause no harm to the sponge, although others are parasites. Many sponges contain single-celled photosynthetic algae (zoochlorellae), blue-green algae, and symbiotic bacteria, which may provide nutrients for the sponge.

Sponges are eaten by sea slugs (nudibranchs), chitons, sea stars (especially in the Antarctic), turtles, and some tropical fishes.

Usually more than half of the species of tropical sponges living exposed rather than under rocks are toxic to fish. This is believed to be an evolutionary response to high-intensity fish predation, nature having selected for noxious and toxic compounds that prevent fish from consuming sponges. Some toxic sponges are very large, such as the gigantic Caribbean sponge (*Cribochalina vasculum*), while others, such as the Caribbean fire sponge (*Neofibularia nolitangere*), are dangerous to the touch—in humans they cause a severe burning sensation lasting for several hours. Toxins probably play an important role in keeping the surface of the sponge clean, by preventing animal larvae and plant spores from settling on them. Some sponge toxins may prevent neighboring invertebrates from overgrowing and smothering them.

Sponge toxins are becoming important in studies on the transmission of nerve impulses. They show considerable potential as biodegradable antifouling agents and possibly as shark repellants.

Bath sponges have been fished in the Mediterranean since very early times. They

Mesozoans

Mesozoans are a taxonomic enigma. They comprise the phylum Mesozoa, which contains about 50 species parasitic on marine invertebrates and none bigger than 0.3in (8mm). They are multicellular animals constructed from two layers of cells, and are therefore distinct from the protozoa, but the layers do not resemble the endoderm and ectoderm of the metazoans (see p12). The features of the group render them unassignable to any other animal phylum. One view holds them to be degenerate flatworms, in other words they may have been previously more complex; the other more widely held opinion, however, is that they are simple multicellular organisms holding a position intermediate between Protozoa and Metazoa.

Mesozoans of the order Dicyemida are all parasites in the kidneys of cephalopods (for example octopus), while the Orthonectida infect echinoderms (starfish, sea urchins etc), mollusks (snails, slugs etc), Annelida (earthworms etc) and ribbon worms (see p63). Despite their simple morphology the Dicyemida have evolved complex life cycles

involving several generations. The first generation, called a nematogen, occurs in immature cephalopods. Repeated similar generations of nematogens are produced asexually by repeated divisions of special central (axial) cells which give rise to worm-like (vermiform) larvae (**1**). When the host attains maturity the parasite assumes the next generation or rhombogen, which looks superficially similar to the nematogen, but differs in its cellular makeup. The individuals are hermaphrodite and produce infusariiform larvae which look superficially like ciliate protozoans. The fate of the larvae is uncertain, but it is believed that another intermediate host is involved in the life cycle. Genera included in the order are *Dicyema*, *Dicyemmerea* and *Conocyema*.

The second of the two orders, the Orthonectida (**2**), live in the tissues and tissue spaces of their marine invertebrate hosts, for example nemerteans, polychaetes, ophiuroids and bivalves. The asexual phase looks like an amoeboid mass and is called a plasmodium because it resembles the protozoan *Plasmodium*.

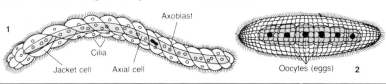

▲ **A blue theme.** These extensive beds of blue sponge encrust the coral limestone around Heron Island near the Great Barrier Reef. The brown lace-like growths are colonies of calcareous bryozoans.

▶ **A squadron of spicules.** Sponge spicules help to support the body of the sponge. They may be made from calcium carbonate or silica. Their various shapes may be characteristic of particular types of sponge and can be important in identifying them. A complex nomenclature has been developed. (**1**) Monaxon spicule with barbs (*Farrea beringiana*). (**2**) Monaxon spicule (*Mycale topsenti*). (**3**) Triaxon spicule (*Leucoria heathi*). (**4**) Hexaxon spicule (*Auloraccus fissuratus*). (**5**) Monaxon spicule with terminal processes (*Mycale topsenti*). (**6**) Monaxon spicule (*Raspaigella dendyi*). (**7**) Monaxon spicule with recurved ends (*Sigmaxinella massalis*). (**8**) Polyaxon spicule (genus *Streptaster*).

owe their usefulness to the water-absorbing and retaining qualities of a complex lattice of spongin fibers; the fibers are also elastic enough to allow water to be squeezed out of the sponge. A number of species are harvested (mainly off Florida and Greece), the chief of which are *Spongia officinalis*, with a fine-meshed skeleton, and *Hippospongia equina*, with a coarser skeleton. These grow on rocky bottoms from low-tide level down to considerable depths and may be collected either by using a grappling hook from a boat, or by divers. The curing of sponges merely involves leaving them to dry in the sun, allowing the soft tissues to rot, pounding and washing them, leaving only the spongin skeleton.

Cultivation of sponges from cuttings has been successfully used although such projects are probably less economic than the making of synthetic products.

Large species, such as the Venus' flower basket have been prized as decorative objects, particularly in Japan.

Sponges contain a variety of antibiotic substances, pigments, unique chemicals such as sterols, toxins, and even anti-inflammatory, and antiarthritis compounds. Boring sponges of the family Clionidae (eg *Cliona* species) may cause economic losses by weakening oyster shells. Boring sponges are widespread within tropical stony corals and cause considerable damage by weakening them. These sponges excavate chambers by both chemical and mechanical methods. GJB

SEA ANEMONES AND JELLYFISHES

Phylum: Cnidaria

About 9,400 species in 3 classes.
Distribution: worldwide, mainly marine; free
swimming and bottom dwelling.
Fossil record: Precambrian (about 600 million
years ago) to present.
Size: microscopic to several yards in width.

Features: radially symmetrical animals with
cells arranged in tissues (tissue grade); possess
tentacles and stinging cells (nematocysts); body
wall of two cell layers (outer ectoderm and
inner endoderm) cemented together by a
primitively noncellular jelly-like mesogloea and
enclosing a digestive (gastrovascular) cavity
not having an anus; there are two distinct life-
history phases: free-swimming medusa and
sedentary polyp.

SEA ANEMONES, corals and jellyfishes are
perhaps the most familiar members of
the phylum Cnidaria. It contains a vast
number of mainly marine animals. There
are only a few freshwater species, of which
the best-known are the hydras.

The cnidarians are multicellular animals
and have a two-layered (diploblastic) construction in which both the differences
between cells and organ development are
limited. These restrictions have, however,
been partially offset in colonial types by the
specialization of individuals (polymorphism). There are two life-history phases:
polyp and medusa. The polyp is the sedentary phase and consists of three regions: a
basal disk or pedal disk which anchors it;
a middle region or column within which is
the tubular digestive chamber (gastrovascular cavity); and an oral region which
is ringed by tentacles. In colonial types a
tubular stolon links adjacent polyps. The
medusa is the mobile phase and is effectively
an inverted polyp. By virtue of the fluid
(water) it contains the digestive cavity plays
an important role in oxygen uptake and
excretion. This fluid additionally acts as a
hydrostatic skeleton through which body
wall muscles can antagonize one another.

Since the medusa is the sexual phase, and
it can be argued that it is the original life
form, with the predominantly bottom-living
(benthic) polyp acting as an intermediate,
multiplicative asexual stage. However, in
the class Hydrozoa the medusa is frequently
reduced or even lost, and in the class
Anthozoa totally absent. Emphasis in the
class Scyphozoa lies, to the contrary, with
the medusa stage, as the evolution of the
highly mobile and graceful jellyfish testifies;
the polyp phase in jellyfish is a relatively
inconspicuous component in the life cycle.

The outer (ectodermal) and inner (endodermal) cell layers of the body are cemented

◄▼ **Some representative species of sea anemones and jellyfishes.** (1) *Cyanea lamarckii*, a jellyfish; N Atlantic (diameter 7.9in, 20cm). (2) *Aurelia aurita*, the Common jellyfish; medusa phase; Mediterranean and N Atlantic (diameter 9.8in, 25cm). (3) *Physalia physalis*, the Portuguese man-of-war; Atlantic (diameter of float 12in, 30cm). (4) *Actinia equina*, the Beadlet anemone; Mediterranean and N Atlantic in the intertidal zone (height 2.8in, 7cm). (5) *Metridium senile*, the Plumose anemone. (6) *Obelia geniculata*, showing a colony of polyps and close-up view; shallow rocky habitats of NW Europe (height of colony 1.6in, 4cm). (7) *Sertularia operculata*, a hydroid; a colony of polyps (height of colony 17.7in, 45cm). (8) *Eunicella verrucosa*, a sea fan; Mediterranean and N Atlantic. (9) *Peachia hastata*, a "sit-and-wait" burrowing anemone; Mediterranean and N Atlantic (length 4in, 10cm). (10) *Corynactis viridis*, an anemone-like animal; N Atlantic (diameter 2in, 5cm). (11) *Alcyonium digitatum*, or dead man's fingers, a colony of polyps; Mediterranean and N Atlantic (height of colony 7.9in, 20cm).

together by the jelly-like mesogloea which in the jellyfish forms the bulk of the animal. The mesogloea contains a matrix of elastic collagen fibers which aid both the change and maintenance of body shape. This is particularly obvious in the pulsating swimming movements characteristic of jellyfish, during which contractions of the swimming bell brought about by radial and circular muscles are counteracted by vertically running, elastic fibers.

Muscle contraction results in an increase in bell depth and hence fiber stress; fiber shortening subsequently restores the bell to its original shape. In the medusae of hydrozoans the resulting water jets are concentrated and directed by the shelf-like velum projecting inwards from the rim of the bell, where there are tentacles, towards the mouth. Structural support in the relatively large anthozoan polyps is also provided by septa (mesenteries) which contain retractor muscles. When mobile, polyp locomotion may be brought about in a number of ways: by creeping upon the pedal disk, by looping or, rarely, by swimming (for example the anemones, *Stomphia*, *Boloceroides*).

The **hydras and their allies** (class Hydrozoa) are considered to be the group that exhibits the most primitive medley of features. The class contains a plethora of medusa and polyp forms which are, for the most part, relatively small. We can plausibly imagine the early hydrozoan life cycle as being similar to that of the hydrozoan order Trachylina. Here the medusae have a relatively simple form and the typical cnidarian

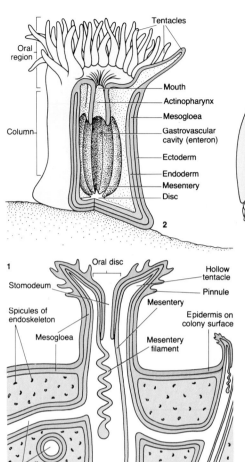

larva, the planula, gives rise in turn to a hydra-like stage which buds-off the next generation of medusae. Significantly this stage is predominantly free-swimming (pelagic), but in other hydrozoan orders subsequent polyp elaboration has resulted in the interpolation of bottom-living, hydra-

◄▲ **The three main forms of cnidarians.** (**1**) A colonial polyp (soft coral). (**2**) A solitary polyp (sea anemone). (**3**) A medusa (jellyfish).

► **Sheltering among the tentacles** of a large anemone, this clownfish (genus *Amphiprion*) probably gains shelter and protection. In return it grooms the anemone's tentacles and brings in scraps of food. Mucus from the anemone probably protects the fish from the stinging cells.

▶ **Paralyzed** OVERLEAF by batteries of stinging cells (shown as fine yellow dots on the tentacles), this worm lies trapped amongst the polyps of a coral (genus *Tubastrea*).

The 3 Classes of the Phylum Cnidaria

Hydras and their allies
Class: Hydrozoa
About 2,700 species in 6 orders (or more).
Fossil record: some hydroids; many hydrocorals.
Features: of 4- (tetramerous) or many- (polymerous) fold symmetry; solitary or colonial; life cycle can include polyp and medusa or exclusively one or other; mesogloea without cells; digestive (gastrovascular) system lacks a stomodeum (gullet); stinging cells (nematocysts) and internal septa absent; sexes separate or individuals bisexual; gametes mature in the ectodermis which frequently secretes a chitinous or calcareous external skeleton; medusa has shelf-to-bell rim (velum); tentacles generally solid.
Orders: Actinulida; Hydroida (hydroids, sea firs); Milleporina; Siphonophora; Stylasterina; Trachylina.

Jellyfishes
Class: Scyphozoa
About 200 species in 5 orders.
Fossil record: minimal.
Features: dominant medusa form with 4-fold (tetramerous) symmetry; polyp phase produces medusae by transverse fission; solitary (either swimming or attached to substrate by stalk); mesogloea partly cellular; digestive (gastrovascular) system has gastric tentacles (no stomodeum) and is usually subdivided by partitions (septa); sexes usually separate; gonads in endodermis; complex marginal sense organs; skeleton absent; tentacles generally solid; exclusively marine.
Orders: Coronatae; Cubomedusae; Rhizostomae; Semaeostomae; Stauromedusae.

Sea anemones, corals
Class: Anthozoa
About 6,500 species in probably 12 orders in 2–4 subclasses.
Fossil record: several thousand species known.
Features: exclusively polyps; predominantly with 6-fold (hexamerous) or 8-fold (octomerous) symmetry; pronounced additional tendency to bilateral symmetry; solitary or colonial; have flattened mouth (oral) disk with an inturned stomodeum; cellular mesogloea; sexes separate or hermaphrodite; gonads in endodermis; digestive (gastrovascular) cavity divided by partitions (septa) bearing gastric filaments; skeleton (when present) is either a calcareous external skeleton or a mesogloeal internal skeleton of either calcareous or horny construction; some forms specialized for brackish water; tentacles generally hollow.

Subclass: Alcyonaria (or Octocorallia)
Orders: Alcyonacea (**soft corals**); Coenothecalia (**blue coral**); Gorgonacea (**horny corals**); Pennatulacea (**sea pens**); Stolonifera; Telestacea.

Subclass: Zoantharia (or Hexacorallia)
Orders: Actiniaria (**anemones**); Antipatharia (**thorny corals**); Ceriantharia; Corallimorpharia; Madreporaria (**hard or stony corals**); Zoanthidea.

Associations and Interdependence in Anemones, Crabs and Fishes

Cnidarians are involved in a variety of associations with other animals, ranging from obtaining food or other benefits from another animal (commensalism) to being interdependent (symbiosis).

An example of commensalism occurs between the hydroid genus *Hydractinia* and hermit crabs, particularly in regions deficient in suitable polyp attachment sites. This is understandable since the shells inhabited by such crabs provide substitute sites and, moreover, the relationship provides *Hydractinia* with the opportunity for scavenging food morsels. Whether any benefit is gained by the crab is unclear, though the development of defensive dactylozooids in *H. echinata* specifically in response to a chemical stimulus emanating from crabs suggests that it does and that there is therefore a mutually beneficial coexistence (mutualism). The association between the cloak anemone *Adamsia palliata* and the crab *Pagurus prideauxi* is, to the contrary, far more intimate. These species normally form a partnership when small. With time the crab comes to outgrow its shelly refuge, but by secreting a horny foot (pedal) membrane the anemone progressively enlarges the shell lip, thus obviating any need for the crab to change shells. A crab lacking an anemone will, upon contact with *Adamsia*, recognize it and attempt to transfer it to its shell.

A range of intermediate degrees of association is provided by the anemone *Calliactis parasitica*, which associates with hermit crabs but also frequently lives independently. An interesting, one-sided association is that of the Hawaiian crab genus *Melia* which has the remarkable habit of carrying an anemone in each of its two chaelae, thereby enhancing its aggressive armament; the crab even raids their food.

Clown fish (*Amphiprion* species) live within the tentacles of sea anemones. Their hosts provide protection to these fish and the fish protect the anemones from would-be predators (primarily other fish which are chased away by the territorial behavior of the clown fish). Additionally the clown fish apparently act as anemone cleaners. Although it has been suggested that inhibitory substances are secreted by these fish to reduce the discharge of stinging cells, recent work has failed to confirm this. It is more likely that stinging is avoided by the secretion of a particularly thick mucous coat, which during acclimatization of the fish to its host anemone may possibly become modified, with the levels of certain excitatory (acidic) components being reduced.

like colonies. Further evolution in specialized niches where dispersal is not at a premium has led in turn to the secondary reduction of medusae; indeed most hydroids lack or almost totally lack a medusoid phase.

Early hydroids were probably solitary inhabitants of soft substrates. Subsequent evolution produced types living in sand (Actinulida) and fresh water (Hydridae—hydras). Most colonial types, however, occur on hard surfaces, anchored by rooting structures. The interconnecting stems (stolons) are protected and supported by a chitinous casing (perisarc) which may or may not enclose the polyp heads. The functional interconnection of members of these colonies permits a division of labor between polyps and an associated variety in form (polymorphism). While one form (gastrozooid) retains both tentacles and a digestive cavity, the form that defends the colony (dactylozooid) has lost the cavity. Another form, gonozooid, is dedicated solely to the budding-off of medusae or, in species lacking medusae, to producing gametes. The delicate branching of hydroid colonies is highly variable, but universally serves to space out member polyps and to raise them well above the substrate, thereby reducing the chance of clogging by silt and sediments.

The evolution of various forms in the class Hydrozoa has culminated in the formation of the complex floating siphonophore colonies (oceanic hydrozoans), each colony composed of a diverse array of both medusae and polyps; they are characteristic of warmer waters. Essentially each individual within the colony is interlinked by a central stolon. In addition to the three polyp forms found in hydroids, there can be up to four forms of medusae: (1) muscular swimming bells that propel the entire colony (for example, *Muggiaea*, *Nectalia*); (2) gas-filled flotation bells (for example, the Portuguese man-of-war—*Physalia*); (3) bracts which play either a supportive or protective role, or both; (4) medusa buds. Freed from the substrate, these colonies are able to reach large sizes with, for example, the trailing colonial stemwork of the Portuguese man-of-war often extending for several yards below the apical float. Such colonies are capable of paralyzing and ingesting relatively large prey items such as fish. Recent research indicates that some species (for example those of the genus *Agalma*) may attract large prey by moving tentacle-like structures, which are replete with stinging cells (nematocysts) and which bear a remarkable resemblance to small zooplankton (copepods).

At one time it was thought that the pinnacle of this evolutionary line was illustrated by animals such as the by-the-wind-sailor (*Vellela*), which has a disk-like, apical float bearing a sail which catches the wind, thus facilitating drifting. It is now thought, however, that these organisms simply consist of one massive polyp floating upside down, and that they are related to gigantic bottom-living hydroids. The large size of these bottom-living giants—up to 4in (10cm) in *Corymorpha* and 10ft (3m) in *Branchiocerianthus*—has been permitted by their adoption of a deposit-feeding life style, often at great depth in still water.

Finally, two groups of hydrozoans produce a calcareous external skeleton: these are the tropical milleporine and stylasterine hydrocorals.

Among cnidarians, it is the **jellyfishes** (class Scyphozoa) that have most fully exploited the free-swimming mode of life though the members of one scyphozoan order (the Stauromedusae) are bottom-living, with an attached, polyp-like existence. Jellyfish medusae have a similar though more complex structure than the medusae of hydroids with the disk around the mouth prolonged into four arms, a digestive system comprising a complex set of radiating canals linking the central portion (stomach) to a peripheral ring element, and a relatively more voluminous mesogloea. The mesogloea in some genera (for example *Aequoria*, *Pelagia*) helps buoyancy by selectively expelling heavy chemical particles (anions) (such as sulfate ions), which are replaced by lighter ones (such as chloride ions). A wide size range of prey organisms are taken, though many species, including the common Atlantic semaeostomes of the genus *Aurelia*, are feeders on floating particles and thus concentrate on small items. The arms of *Aurelia* periodically sweep around the rim of the bell, gathering up particles which accumulate there following deposition on the animal's upper surface. In contrast the arms around the main mouth of the Rhizostomae have become branched, and have numerous sucking mouths, each capable of ingesting small planktonic organisms such as copepods. Within this group are the essentially bottom-living, suspension-feeding forms of the genus *Cassiopeia*, which lies upside down on sandy bottoms, their frilly arms acting as strainers. The bell shapes of members of two orders are distinctive: coronate medusae have bells with a deep groove and cubomedusae have bells that are cuboid in shape.

The gametes of jellyfish are produced in gonads which lie on the floor of the digestive cavity and are initially discharged into it. Fertilization normally occurs after discharge of the gametes. Many species, however, have so-called brood pouches located on the undersurface where the larvae are retained. After release larvae settle and give rise to polyps which produce additional polyps by budding. These polyps also produce medusae by transverse division (fission), a process which results in the formation of stacks (strobilae) of so-called ephyra larvae. When released the ephyra larvae feed mainly on protozoans and grow and change into the typical jellyfish.

Corals and sea anemones (class Anthozoa) only exist as polyps. Sea anemones (order Actiniaria) always bear more than eight tentacles and usually have both tentacles and internal partitions (mesentaries) arrayed in multiples of six.

Many anemone species, especially the more primitive ones, are burrowers in muds and sands but most dwell on hard substrates, cemented there (permanently or temporarily) by secretions from a well-differentiated disk. The disk around the mouth (oral disk) is provided with two grooves (siphonoglyphs) richly endowed with cilia which serve to maintain a water flow through the relatively extensive, digestive cavity. The oral disk extends inwards to produce a tubular gullet or stomodeum which acts as a valve, closing in response to increases in internal pressure. In common with jellyfish some anemones feed on particles suspended in the seawater for which leaf-like tentacles, prodigious mucus production and abundant food tracts lined with cilia are required; a good example is the common plumose anemone, genus *Metridium*. Asexual reproduction occurs by budding, breaking-up or fission, while sexual reproduction may involve either internal or external fertilization of gametes. Some species brood young, either internally or externally at the base of the column.

Members of two other orders are also anemone-like: the cerianthids have greatly elongated bodies adapted for burrowing into sand, but have only one oral groove (siphonoglyp). Zoanthids lack a pedal disk, are frequently colonial and often live attached to other organisms (epizoic).

Also included in the subclass Zoantharia are the hard (stony) corals (order Madreporaria) whose polyps are encased in a rigid, calcium carbonate skeleton. The great majority of hard corals live in colonies

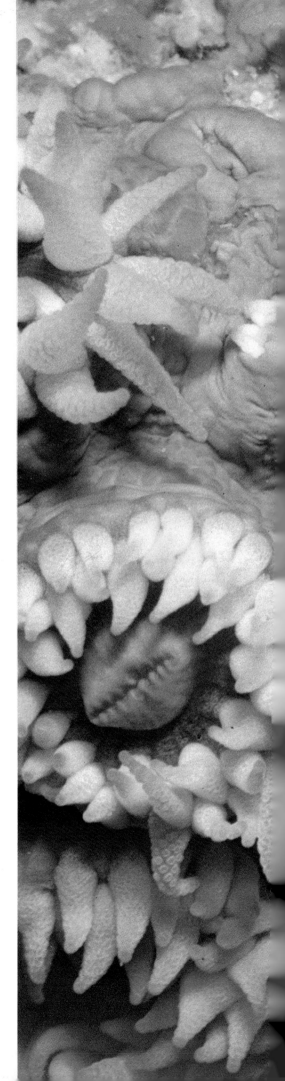

which are composed of vast numbers of small polyps (about 0.2in, 5mm), but the less abundant solitary forms may be large (*Fungia* up to 20in, 50cm, across); most are tropical or subtropical in distribution. In colonial forms the polyps are interconnected laterally; they form a superficial living sheet overlying the skeleton, which is itself secreted from the lower outer (ectodermal) layer.

Corals exhibit a great diversity of growth forms, ranging from delicately branching species to those whose massive skeletal deposits form the building blocks of coral reefs (see pp44–45). An interesting growth variant is exhibited by *Meandrina* and its relatives in which polyps are arranged continuously in rows, resulting in the production of a skeleton with longitudinal fissures, a feature which accounts for its popular name, the brain coral.

Closely related to the hard corals are the members of the order Corallimorpharia which lack a skeleton. Included in this group is the jewel anemone (genus *Corynactis*), so named because of its vivid and highly variable coloration. Since it reproduces asexually, rock faces can become covered by a multicolored quiltwork of anemones. The black or thorny corals (order Antipatharia) form slender, plant-like colonies bearing polyps arranged around a horny axial skeleton; they possess numerous thorns.

Octocorallian corals comprise a varied assemblage of forms, but all possess eight feather-like (pinnate) tentacles. The polyps project above and are linked together by a mass of skeletal tissue called coenenchyme, which consists of mesogloea permeated by digestive tubules. Thus in contrast to hard corals, the octocorallians have an internal skeleton. This assemblage includes the familiar gorgonian (horny) corals, sea whips and fans, and the precious red coral, genus *Corallium*. Most of these have a central rod composed of organic material (gorgonin) around which is draped the coenenchyme and polyps, the former frequently containing spicules which may impart a vivid coloration. Such is the case with *Corallium*, whose central axis consists of a fused mass of deep red calcareous spicules; this material

is used in jewelry. The tropical organ pipe coral, genus *Tubipora* (order Stolonifera) produces tubes or tubules of fused spicules which are cross-connected by a regular series of transverse bars. In contrast, the soft corals (order Alcyonacea) only contain discrete spicules within the coenenchyme (for example dead man's fingers, genus *Alcyonium*). The order Coenothecalia is solely represented by the Indo-Pacific blue coral, genus *Heliopora*, which has a massive skeleton composed of crystalline aragonite

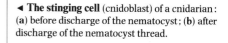

▲ **A waving mass of tentacles**—this dahlia anemone (*Tealia felina*) traps prey 65ft (20m) below the surface of the North Sea. In the foreground are the arms of a brittle star.

◄ **The stinging cell** (cnidoblast) of a cnidarian: (**a**) before discharge of the nematocyst; (**b**) after discharge of the nematocyst thread.

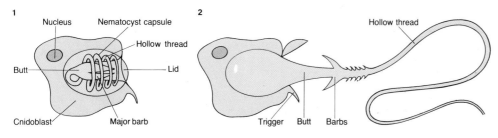

fibers fused into plates (lamellae): its blue color is imparted by bile salts. Many species in most of these groups have several forms (especially gastrozooids, dactylozooids and gonozooids). This is also true of the sea pens (order Pennatulacea) which are inhabitants of soft bottoms. Each possesses a large, stem-like, primary polyp (as is also the case in the order Telestacea) housing a skeletal rod which becomes embedded in the substrate as a result of waves of contractions. Secondary polyps are arranged laterally on this stem and exhibit two forms; many polyps (siphonozooids) act as pumps, promoting water circulation through the colonial digestive system. Familiar examples are the sea pansy (genus *Renilla*) and the sea pen (genus *Veretillum*), both of which when disturbed exhibit waves of glowing phosphorescence. These are controlled by the nervous system and are inhibited by light. Their role is not clear, though it is likely that they are a response to intrusion by would-be predators such as fish. A number of other cnidarians display a similar phenomenon, a good example being the hydromedusae of the genus *Aequorea*.

To trap prey cnidarians normally employ stinging cells (nematocysts). The discharge of these is now thought to be under nervous control. Discharge involves a collagenous thread being rapidly shot out, uncoiling and turning inside out in the process, sometimes to expose lateral barbs. Hollow stinging cells frequently contain a toxin which can enter the body of the prey. The released toxins may be extremely potent: especially dangerous are those released by the cubomedusan sea wasps (for example, genus *Chironex*)—

jellyfish that have been responsible for killing several humans, particularly off Australian coasts. Victims usually succumb rapidly to respiratory paralysis. Nematocysts may be pirated by sea slugs and used for their own protection (see p127).

The cnidarian nervous system shows a certain amount of organization and local specialization. This is especially evident in anemones where nerve tracts accompany the retractor muscles responsible for protective withdrawals. The marginal ganglia of scyphomedusae and the circumferential tracts of hydromedusae have been found to contain pacemaker cells which are responsible for initiating and maintaining swimming rhythms. In *Polyorchis* it has been found that the giant nerves controlling movement are all coupled together electrically, ensuring that they function collectively as a giant ring nerve fiber capable of initiating synchronously muscle contraction from all parts of the bell.

Similarly the behaviors of individual polyps in hydroid and coral colonies are integrated by the activities of colonial nerve nets. Additional powers for integrating control are provided by conduction pathways apparently constituted by sheets of electrically coupled, epithelial cells. For example, the shell-climbing behavior of anemones of the genus *Calliactis* seems to depend upon the interplay of activities between two epithelial systems—one on the outside (ectodermal), the other inside (endodermal)—and the nerve net, though conclusive evidence as to the exact cellular locations of these additional systems has been difficult to obtain.   RB

Aggression in Anemones

A number of anemones display a well-defined aggressive sequence which, for the most part, is used in confrontation with other anemones. These anemones all possess discrete structures located at the top of the column which contain densely packed batteries of stinging cells (nematocysts): they are called acrorhagi. They can be inflated and directed at opponents.

The common intertidal beadlet anemone (*Actinia equina*), whose distribution encompasses the Atlantic seaboards of Europe and Africa and which also occurs in the Mediterranean, is an example upon which attention has been focused. Although this species can vary considerably in color (red to green), the acrorhagi are always conspicuous thanks to their intense bluish hue. Aggression is triggered by the contact of tentacles. One individual usually displays column extension and bends so that some of its simultaneously

enlarging acrorhagi make contact with the opponent (after 5–10 minutes). There follows a discharge of the stinging cells (nematocysts) which normally results in a rapid withdrawal by the victim.

Experiments have suggested that in common with more advanced animals contest behavior is ritualized, but that in these lowly forms it is dependent upon "simple physiological rules" rather than upon complex behavioral ones. The "rules" used apparently decree that larger anemones should act aggressively more rapidly than smaller ones and, as a result, will subsequently win contests.

The North American anemone *Anthopleura elegantissima* reproduces asexually by fission. In consequence intertidal rocks can become entirely covered by a patchwork of asexually produced anemones. Close inspection reveals that each densely packed clone (ie the mass of

asexually produced offspring) is separated from its neighbors by anemone-free strips, and that these are maintained by aggressive interactions involving acrorhagi. It is clear, therefore, that the aggressive behavior of individuals constituting the boundary of a clone serves to provide territorial defense for the entire clone, the central members of which, significantly, are more concerned with reproduction than aggression. Thus there is, as in hydroid colonies, a functional division of labor, despite the lack of physical interconnection between the clonal units. Individuals at the interclonal border have more and larger acrorhagi than centrally placed members, a difference apparently dependent solely on the former experiencing aggressive contact with nonclonemates. Such a dichotomy is thought to be indicative of the presence of a sophisticated self/nonself recognition system.

The Living and the Dead

The origins and biological organization of coral reefs

Coral reefs are extraordinary oases in the midst of oceanic deserts, for they support immensely rich and diverse faunas and floras but occur primarily in the tropics where the marked clarity of the water indicates a relative dearth of planktonic organisms and other nutrients. "Coral" consists of the skeletons of hard or stony coral.

The success of reef-building (hermatypic) corals in tropical waters, where high light intensities prevail throughout the year, is strictly dependent upon the nutrient-manufacturing (autotrophic) activities of interdependent (symbiotic) algae (zooxanthallae) which live within each polyp. Such dependence also necessarily restricts the algae to these waters. Moreover, since these algae flourish best at temperatures higher than 68°F (20°C), reef development is further limited to depths of less than 230ft (70m), ie where light intensities are greatest. Corals do, however, survive both at higher latitudes and in deeper waters, but where they do their capacity to secrete limestone for reef building is found to be severely curtailed as a result of the reduced metabolic support provided by the algae. Finally, restrictions on the distribution of corals are also imposed by the deposition of silt, freshwater run-off from land and cold, deepwater upwelling. The two former factors, for example, restrict reef development in the Indo-Pacific Ocean towards offshore island sites, while the last hinders coral growth off the west equatorial coast of Africa, where the Guinea current surfaces. It should not be forgotten, though, that in the development of most reefs, encrusting (calcareous) algae (for example, *Lithophyllum*, *Lithothamnion*) normally play an extremely important part. In many cases the limestone they produce acts as a cement.

In 1842 Charles Darwin distinguished three main geomorphological categories of reef which are still in use today: fringing reefs, barrier reefs and atolls. As the name suggests, the first are formed close to shore, on rocky coastlines. Barrier reefs are, on the contrary, separated from land by lagoons or channels which have usually been produced as a result of subsidence. (The best known and largest barrier reef is the Great Barrier Reef off the northeast coast of Australia, though the name is somewhat misleading as along its length (1,200mi, 1,900km) occur a host of different reef configurations—more than 2,500). Finally, atolls are found around subsiding volcanoes.

The continuation of coral growth is heavily dependent upon changes in water level. At present the world is in a period between glaciations and the rising sea level permits vertical growth to continue at about 0.1–0.6in (0.3–1.5cm) per year, a rate that has apparently been maintained over the last 100,000 years. Core drillings taken at Eniwetok atoll in the Pacific have extended downwards for up to 1mi (1.6km) before hitting bed rock: from analyses of both the fauna and flora in cores obtained, it has proved possible to reconstruct past fluctuations in sea level. The majority of the world's coral reefs started development during the Cenozoic era (not later than 65 million years ago) and consist predominantly of corals of the order Madreporaria.

All coral reefs have a similar biological organization with the reef plants and animals, as on rocky shores, lying in zones in accordance with their tolerances to physical factors. This is most evident on the exposed, windward faces of those reefs subject to continuous wave crash where especially prolific growths of both corals and algae develop. However, a reef can only grow outwards if debris accumulates on the reef slope; with increasing water depths, such material tends to slide down the slope and thus becomes unavailable. Below 100–165ft (30–50m), hermatypic corals are replaced by nonhermatypic ones, and by fragile, branching alcyonarians (gorgonians, etc). Above the slope is the reef crest which, in the most exposed situations, is dominated by encrusting calcareous algae (for example, the genus *Porolithon*). These algae form ridges which are full of cavities thereby providing numerous recesses which

▲ **Life around Bermuda**, a coral reef scene from the western Atlantic. The dominant organism here is the Common sea fan (*Gorgonia ventilina*), which is surrounded by hard brain corals and erect sponges.

◄ **Life on a coral surface.** Coral limestone and the skeletons of dead corals provide a substrate for many small encrusting organisms, including other cnidarians. Here *Parazoanthus swiftii* (not itself a coral) is spreading its colonial polyps to capture food suspended in the water.

are colonized by a multitude of invertebrates, including zoanthids, sea urchins and vermetid gastropods. Where wave action is not too severe, windward reef crests are usually dominated by a relatively small number of coral species, notably stout *Acropora* and hydrocoralline *Millepora* species.

Zonation is far less marked on the leeward side of the reef crest, where a different set of problems has to be faced of which sediment accumulation from land run-off is perhaps the most acute. Nevertheless, the relatively sheltered nature of this habitat permits the rapid proliferation of branched corals. In common with the coral faunas of

windward slopes, those of leeward faces display dramatic changes of coral form with increasing depth, changes which can either be attributable to the replacement of species by others or to changes in species forms. For example, on Caribbean reefs, dominant species such as *Monastrea annularis* display both stout (shallow-water) and branching (deep-water) growth forms. Recent work on *M. cavernosa* has indicated that forms found at equivalent depths are distinctive: the polyps of the shallow-water form are open continuously, whilst those of the deep-water form (which house far fewer zooxanthellae) are open only at night. RB

COMB JELLIES

Phylum: Ctenophora

About 100 species in 5 orders (sometimes grouped into 2 classes).

Distribution: worldwide, marine.

Fossil record: none.

Size: from very small (about 0.15in, 0.4cm) to over 3.3ft (1m) in length.

Features: basically radially symmetrical but masked by superimposed bilateral symmetry; body wall 2-layered (diploblastic) with a thick jelly-like mesogloea and nerve net (these features making them similar to sea anemones and jellyfishes); 8 rows of plates of fused cilia (comb plates) upon whose activity locomotion predominantly depends; tentacles when present help in the capturing of zooplankton; digestive/gastrovascular system with a stomodeum (gullet), stomach and a complex array of canals; one phase in life cycle (not equivalent to polyp or medusa); bisexual.

Orders: Beroida (class Nuda), including *Beroë gracilis*; Cestida, including **Venus's girdle** (genera *Cestum*, *Velamen*); Cydippida, including *Pleurobrachia*; Lobata, including *Mnemiopsis*; Platyctenea, including *Coeloplana*, *Tjalfiella*, *Ctenoplana*, *Gastrodes*.

▼ **Feeding time in the oceans:** a comb jelly (genus *Lampea*) eating a chain of red salps. The comb jelly's coiled white tentacle and the rows of cilia (comb plates) on the ridges of the animal can be seen in this picture.

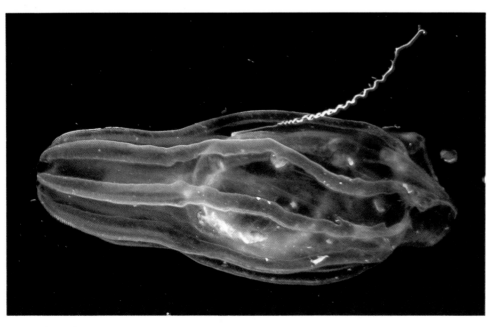

SMALL, translucent, gelatinous globular animals, comb jellies float through the open seas, like ghosts, capturing prey with their whip-like tentacles. The body consists of three zones—a voluminous mesogloea sandwiched, as in sea anemones and jellyfishes, between thin ectodermal (outer) and endodermal (inner) cell layers. Most noticeable, however, are the eight rows of plates of cilia (comb rows) whose activities serve to propel the animal while it is searching for zooplanktonic prey, which it captures by means of the pair of tentacles loaded with adhesive cells.

Most comb jellies resemble species in the common genus *Pleurobrachia* (order Cydippida) which occurs in the colder waters of both the Atlantic and Pacific oceans, and is often found stranded in tidal pools. Its globular body is up to 1.5in (4cm) in diameter, and has two pits into which the tentacles can be retracted. These tentacles function as drift nets, catching passing food items while the animal hovers motionless. When extended these appendages may be up to 20in (50cm) in length. They have lateral filaments and bear numerous adhesive cells (colloblasts), each of which has a hemispherical head fastened to the core of the tentacle by a straight connective fiber and by a contractile spiral one, the latter acting as a lasso. Once caught prey is held by the colloblasts, which produce a sticky secretion, until transferred to the central portion (stomach) of the digestive system following the wiping of the tentacles over the mouth. It has been reported that when feeding upon pipe fishes of a similar size to itself, *Pleurobrachia* will play them in much the same way as an angler tires out a hooked fish. The stomach, in which digestion commences, leads to a complex array of canals where there is further digestion and intake of small food particles which are subsequently broken down by intracellular digestion. These canals are especially routed alongside those body regions having high energy consumption levels, notably the eight comb rows. The gonads lie in association with the lining of the gastric system; gametes pass out through the mouth. Following external fertilization a larva, which is a miniature version of the adult, is produced.

The common comb jellies of temperate waters are all like *Pleurobrachia*, which has the most primitive body form: shape in the other orders departs from this. The elongate lobate comb jellies are laterally compressed and have six lobes projecting from their narrow mid region, four of which are delicate and two stout. These serve to capture food: the tentacles are small and lack sheaths. *Mnemiopsis*, which is about 1.2in (3cm) in size and occurs in immense swarms, has, in association with the production of these lobes, four long and four short comb rows. Elongation and compression have been carried much further in the Cestida, resulting in organisms resembling thin gelatinous bands. Species in the two genera concerned (*Cestum*, *Velamen*) are collectively referred to as Venus's girdle, and are found in tropical waters and the Mediterranean, only occasionally straying into northern waters. The graceful swimming of these forms is principally dependent upon undulations of the whole body, brought about by muscle fibers embedded in the mesogloea. They feed entirely by means of tentacles set in grooves running along the oral edge. In the order Beroidea, the thimble-shape body is similarly laterally compressed, but mainly occupied by the greatly enlarged stomodeum rather than mesogloea. Species in this group are up to 8in (20cm) in length and often have a pinkish color. There are no tentacles; instead food is caught by lips which curl outwards to reveal a glandular and ciliated (macrocilia) area. Relatively large food items are rapidly taken in by a combination of suction pressure (brought about by the contraction of radial muscles in the mesogloea) and ciliary action. The common North Sea species *Beroë gracilis* feeds exclusively upon *Pleurobrachia pileus*.

The final order, the Platyctenea, is a curious group which, contrary to other flattened ctenophores, are compressed from top to bottom and have, for the most part, assumed a bottom-living (benthic) creeping

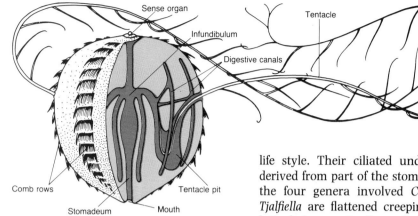

Sense organ

Tentacle

Infundibulum

Digestive canals

Comb rows

Tentacle pit

Stomadeum

Mouth

▶ **The structure of a comb jelly** (*Hormiphora plumosa*).

▼ **A gentle predator.** A comb jelly (genus *Pleurobrachia*) trails its adhesive tentacles—like fishing lines—through the water to ensnare microscopic food organisms. Comb jellies are noted for their luminescence, which is produced by glandular structures—appearing as greenish streaks—lying in association with their eight radial gastrovascular canals.

life style. Their ciliated under surface is derived from part of the stomodeal wall. Of the four genera involved *Coeloplana* and *Tjalfiella* are flattened creeping forms, the latter being practically sessile (ie attached to the substrate). In contrast *Ctenoplana* is partially planktonic, having become adapted to creeping on the water's surface also. The most specialized, though, is *Gastrodes* which is a parasite of the free-swimming sea squirt genus *Salpa*. The flattened adult stage is a free-living, bottom-dwelling form, but the planula-type larva bores into the tunicate test and develops into a bowl-shaped, intermediate parasitic stage.

The waves of movement of the comb plates responsible for swimming are initiated and synchronized by impulses arising within the apical sense organ which functions primarily as a statocyst, detecting tilting, although it is also sensitive to light. It contains a sensory epithelium bearing four groups of elongate sensory "balancer" cilia, upon which a calcareous ball (statolith) perches. The orientation of the animal is controlled very simply: irrespective of whether the comb jelly is swimming upward or downward, deflection of the balancer cilia away from an upright position brought about by the statolith results in a change in the frequency of beating of the cilia which is spontaneously generated. The overall effect exerted through the differential activitation of the four groups of "balancer cilia," and subsequently the four pairs of comb rows to which they are electrically connected, is to ensure that the animal swims vertically rather than obliquely. Since the power stroke of the individual comb plates is opposite to direction of waves of activity passing along the comb rows, the animal normally travels mouth forward. However, this activity is also under control exerted by the nerve net, which upon receipt of a mechanical stimulus can cause either a cessation or reversal of beating. There is also some evidence to suggest that the synchronization of beating of the comb plates within any one row might be partially dependent upon direct mechanical coupling.   RB

ENDOPROCTS

Phylum: Endoprocta (or Entoprocta or Kamptozoa)
Over 130 species in 3 families.
Distribution: worldwide, mainly marine.
Fossil record: none.
Size: minute, 0.02–0.16in (0.5–4mm),

Features: stalked, body cavity a pseudocel; solitary or colonial; body comprises a stalk or peduncle supporting a calyx bearing a circle of partially retractile tentacles within which both mouth and anus open; the calyx encloses paired excretory protonephridia with flame cells; no vascular system; sexes on some or different individuals depending on species; sexual and asexual reproduction.

Family: Pedicellinidae, including *Pedicellina*, *Myosoma*, *Barentsia*, and 3 other genera.

Family: Urnatellidae, sole genus *Urtanella*.

Family: Loxosomatidae, including *Loxosoma*, *Loxocalyx*, *Loxosomella*, *Loxomespilon*, *Loxostemma*.

ENDOPROCTS are small inconspicuous animals that are usually less than 0.04in (1mm) in length. Most inhabit the sea but one genus lives in fresh water. They are sedentary and live attached to hard surfaces or other organisms by a stalk. In the latter case they live as commensals neither receiving benefit from nor giving benefit to the host. Some of the commensal species, such as *Loxosomella phascolosomata*, are catholic and dwell on various marine invertebrates such as the sipunculids *Golfingia* and *Phascolion* as well as on some bivalve shells. Others, like *Loxomespilon perenzi*, are host-specific living on the polychaete worm *Sthenelais*. Some are even specific to certain parts of the host like the exhalent pores or oscula of sponges or the segmental appendages or parapodia of polychaete worms. Preferences for these specific habitats among endoprocts may reduce competition for survival between endoproct species and place individuals where they can most benefit from water currents etc.

Relatively little is known about endoprocts. The European fauna is best known and it is certain that many species remain to be discovered, particularly in the tropics.

Individual endoprocts are called zooids and live, according to species, either as solitary animals or in a colony where many zooids are linked together by creeping root-like stolons. The zooids have a fairly simple structure, each one consisting of a cup-like body called the calyx supported on a stalk-like peduncle. The calyx and peduncle are comprised of body wall tissue, which is soft and flexible, cloaked with a thin protective layer of cuticle under which lies a thin layer of epidermis and a thin layer of muscle. The muscle brings about the nodding movements of the calyx on the peduncle. Towards its top the calyx is slightly constricted by a rim above which radiate about 40 hollow

▶ **The structure of endoprocts.** (1) A section showing in detail the internal structure of a zooid (genus *Pedicillina*). (2) Part of a colony of *Pedicillina* showing each zooid linked by a creeping stolon. Many endoprocts live in association with other host animals, from whose feeding and respiration currents individual endoproct zooids obtain their own nutrient requirements. (3a) One such host is the sea mouse (*Aphrodite aculeata*), upon which lives the endoproct *Loxosemella fauveli*, individuals of which are shown (3b) after removal of the scales covering the body of the sea mouse.

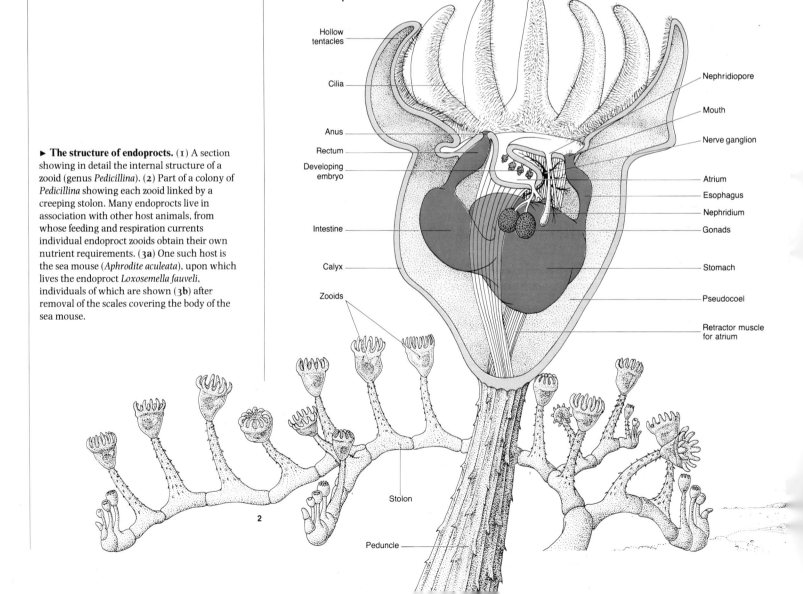

1

Hollow tentacles
Cilia
Anus
Rectum
Developing embryo
Intestine
Calyx
Zooids

Nephridiopore
Mouth
Nerve ganglion
Atrium
Esophagus
Nephridium
Gonads
Stomach
Pseudocoel
Retractor muscle for atrium

2

Stolon
Peduncle

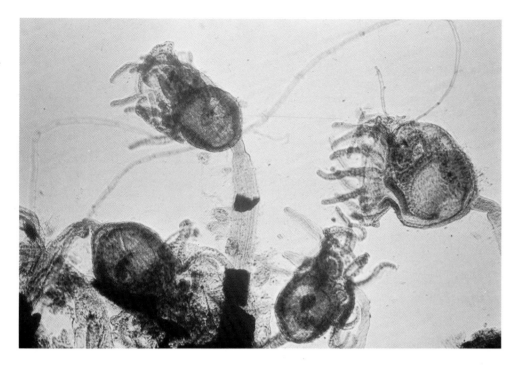

▲ **Animals on stalks,** zooids of the endoproct genus *Pedicellina*.

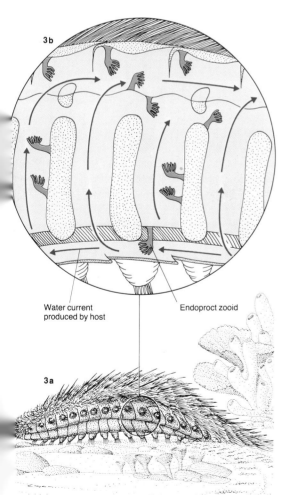

Water current produced by host

Endoproct zooid

water current leaving the tentacle crown.

Many species undergo asexual reproduction by budding. The style of budding, for example directly from the calyx or from the peduncle (as well as the pattern of growth form in colonial types), is characteristic of particular genera. Budding from the calyx is customary in solitary forms like *Loxosomella*. When the buds reach an advanced stage they drop off to occupy a new site and lead an independent life.

Some species are hermaphrodite, others separate-sexed. Fertilization is believed to be within each zooid. The resulting embryos are then brooded in the atrium until they are released as free-swimming larvae. Following the planktonic phase the larvae settle and grow into new zooids.

The three families of endoproct (Pedicellinidae, Urnatellidae and Loxosomatidae) can be distinguished by the form of the zooids and the growth habit. In the first family there are no solitary species. Here the calyx is separated from the peduncle by a diaphragm so that when conditions are unfavorable or when damaged by predators, such as small grazing arthropods, the calyx can be shed. It can be regrown: examples of this family are frequently met with showing a range of regenerating calyces.

The Urtanellidae contains only one genus, *Urtanella*, which is the only fresh-water one. Here the stolon is small and disk-like and the calyces may be shed and show regeneration frequently. One species occurs in both western Europe and the eastern USA; a second is known from India.

The Loxosomatidae, which are the most abundant endoproct species, are all solitary, usually with a short peduncle attached by a broad base with a cement gland or a muscular attachment disk. The former type cannot move, being cemented down, but the latter can detach and reattach themselves as conditions require. The peduncle and calyx are continuous and there is no diaphragm. The calyx cannot be shed in the Loxosomatidae.

For many years endoprocts were classified with the ectoprocts (moss animals or sea mats) in one phylum, the Polyzoa, but most authorities now regard them as a separate group. One key difference between the two is that in ectoprocts (phylum Bryozoa or Ectoprocta) the anus opens outside the ring of tentacles, not within the calyx. In the endoprocts the anus (proctodeum in anatomical terms) lies within the ring of tentacles. This explains the meaning of the two names: Endoprocta, anus inside; Ectoprocta, anus outside. AC

tentacles. These can be folded over the top of the calyx and partially retracted so that the muscular web of tissue which connects their bases affords them some protection in unfavorable circumstances. The tentacles are covered with cilia, minute beating threads projecting from the skin.

On top, inside the circlet of tentacles, the calyx is penetrated by the mouth and anus, between which runs a U-shaped gut. The space inside the calyx is taken up by the body cavity (pseudocoel), the fluid contents of which bathe the gut and other internal organs. The body cavity fluid contains some free cells which wander about. The reproductive organs lie close to the gut and their short ducts discharge into a fold, the atrium, in the top of the calyx; embryos may be brooded in the atrium. Endoprocts lack an internal circulatory system and have no special respiratory structures; oxygen dissolved in the surrounding water simply diffuses into the zooid. A pair of nephridia are responsible for excretion of nitrogenous waste and these discharge through a single nephridiophore on the top of the calyx just behind the mouth.

Endoprocts are suspension feeders utilizing small organic particles and microorganisms borne in the water currents to supply them with food. The cilia on the tentacles drive water in between the tentacle bases and up and out through the central opening in the tentacle crown. The food particles are caught by the cilia on the sides of the tentacles and passed down to the mouth bound up in a string of mucus. Unwanted particles are flicked into the

Phyla: Rotifera, Kinorhyncha, Gastrotricha

Rotifers or wheel animalcules

Phylum: Rotifera
About 1,700 species in about 100 genera.
Distribution: worldwide, mainly freshwater
and damp soils.
Fossil record: none.
Size: microscopic, 0.002–0.08in (0.04–2mm).

Features: solitary or colonial; body cavity a
pseudocoel; body comprises head section, trunk
and tail piece; head bears a crown of hairs
(cilia); trunk houses internal jaws (mastax);
tailpiece in some forms bears gripping toes;
excretion by means of flame cell
protonephridia; body sometimes encased in a
lorica.

Kinorhynchs

Phylum: Kinorhyncha
About 120 species in the class Echinoderida.
Distribution: probably worldwide in marine
coastal areas.
Fossil record: none.
Size: less than 0.04in (1mm).

Features: free-living; body segmented and
covered in spines (no cilia); excretion via a pair
of protonephridia, each fed by a single flame
cell.

Gastrotrichs

Phylum: Gastrotricha
About 200 species.
Distribution: widespread, marine and fresh
water.
Fossil record: none.
Size: microscopic.

Features: free-living usually with some external
areas covered with cilia; body cavity a
pseudocoel; body of three layers (triploblastic)
often with one or more pairs of adhesive
organs; excretory protonephridia when present
consist of a single flame cell; cuticle covered
with spines, scales or bristles.

Class: Chaetonotoida
Mainly fresh water. Front end of body usually
distinct from trunk; adhesive tubes at rear;
protonephridia present; mainly
parthenogenetic (asexually reproducing)
females, including genus *Chaetonotus*.

Class: Macrodasyoidea
Marine, chiefly reported from Europe. Bodies
straight with adhesive tubes on front, rear and
sides of the body; no protonephridia; bisexual;
including genus *Macrodasys*.

▶ **Little animals of great endurance.** ABOVE
Antarctic rotifers (*Philodina gregeria*) have
survived being frozen for over a hundred years.
They can also withstand immersion in liquid
nitrogen.

▶ **The structures of representative genera** of
(1) rotifers (genus *Brachionus*); (2) kinorhynchs
(genus *Echinoderes*); (3) gastrotrichs (genus
Chaetonotus).

ROTIFERS' most conspicuous structure is
their crown of hairs (cilia) borne on a
retractable disk: the corona. The cilia beat
in such a way that it resembles a wheel spinning, or two wheels spinning in opposite
directions when viewed under the microscope. Thus the early microscopists termed
rotifers "wheel animalcules."

The three phyla treated here are relatively
little-known members of the animal kingdom. They are all aquatic, living mainly in
fresh water. Some authorities have grouped
them with the roundworms and horsehair
worms in one phylum, known as the
Aschelminthes, on account of certain developmental and structural similarities. Others
believe that they are best considered phyla
in their own right because the criteria for
grouping them are somewhat debatable.

All these animals show bilateral symmetry and a body cavity, but the latter is a
pseudocoel lying between the body wall and
the gut (see p12) and not a true coelom.
The animals are not segmented and the
body is supported by a layer of skin (cuticle).
The alimentary tract is usually "straight
through," with a mouth, esophagus,
intestine, rectum and anus. Excretion is
carried out through structures known as
protonephridia. Reproductive strategies
vary from group to group although the sexes
are usually separate. Development is along
protostome lines (see p16).

The **rotifers** constitute quite a large phylum
with many species living in fresh water and
soils all over the world. They inhabit lakes,
ponds, rivers and ditches as well as gutters,
puddles, the leaf axils of mosses and higher
plants and damp soil. Most are free living,
but a few form colonies and some live as
parasites.

Rotifer bodies can generally be divided
into a head at the front, a long middle section or trunk housing most of the viscera,
and a tailpiece or foot terminating in a gripping toe. The head as such is not well developed in comparison with those of the higher
animals but it does bear a characteristic eye
spot sometimes colored red. The most conspicuous structure is the corona. The beating of its cilia draws in a stream of water
toward the head which brings with it food
in the form of other microorganisms. These
are then passed into the muscular esophagus which houses powerful teeth and jaws.
The chewing structure is technically termed
the mastax and the teeth and jaws are made
of strong cuticle for chewing and grinding
the food. In some predatory rotifers long
teeth may protrude out of the mouth and

be used to seize protozoans and other species
of rotifer.

The trunk houses the internal organs
including the mastax. Its contents appear
complex: they include elements of the gut,
excretory system and reproductive system
as well as the musculature for retracting the
corona. The tailpiece is important for posture and stance—many rotifers use it to
cling to fronds of aquatic plants or other
substrates while they feed or rest. The tail
may also serve as a sort of rudder.

The breeding strategies of rotifers typically include a phase of reproduction where
females lay unfertilized eggs which develop
and grow into other females (parthenogenetic). When the sexual season dictates or
when adverse environmental conditions
prevail some of the females lay eggs which
need to be fertilized by a male. At the same
time others lay smaller eggs which develop
into males for this purpose. The males then
fertilize the new females by injecting them
with sperm through the body wall. The
resulting fertilized eggs are tolerant of harsh
conditions and can withstand drought, an

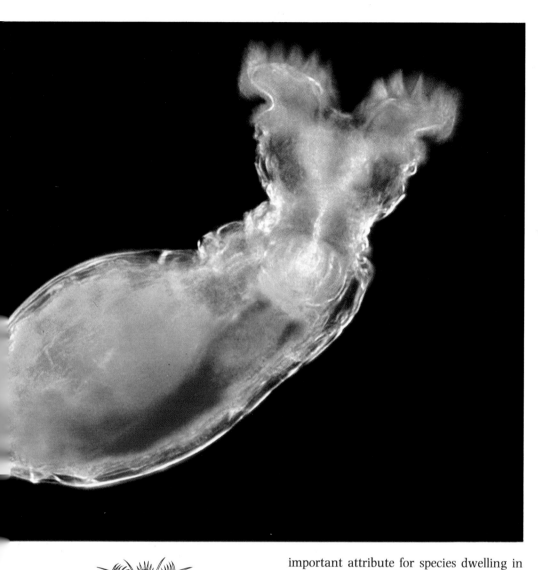

Despite their obscurity rotifers are of ecological and direct economic importance. Millions of them may exist in a small body of water and their populations may rise and fall rapidly. They are important members of numerous aquatic food webs. Soil-dwelling rotifers play a role in soil breakdown.

The **kinorhynchs** are poorly known marine "worms" reaching up to 0.04in (1mm) in length. They probably exist worldwide but most of the 100 or so known species have been recorded from European coastal sands and muds—a reflection probably of where people have searched for them rather than of their actual distribution. Their bodies appear segmented but the divisions are only "skin-deep" and they are quite unrelated to the annelid worms or arthropods. They creep about in sand and mud using the spines of the head to gain a hold while the rear of the body is contracted forward. Then the tail spines dig in and the head is advanced and so the process is repeated. The head is not well developed. Male and female kinorhynchs have a similar appearance and sexual reproduction occurs year round. A larva emerges at hatching which molts several times before becoming an adult.

Gastrotrichs are also minute animals often found in habitats shared with rotifers. They are "worm-like" and live among detrital particles in both fresh and seawater. The genus *Chaetonotus* is a good example. Its head is rounded and attached to the trunk by an elongated neck; the tail is usually forked and the outer surfaces of the body are covered with sticky knobs (papillae). The underside bears hairs (cilia) arranged in bands whose coordinated beating causes the animal to glide over the substrate. The upper surface of the animal is usually armed with spines and scales.

The gastrotrich pharynx generates a sucking action which is employed to draw food into the gut. It takes single-celled algae, bacteria and protozoans. The reproductive strategies vary according to the groups of gastrotrichs. Males are unknown in the class Chaetonotoidea where reproduction is by parthenogenesis among females. Here two types of egg are laid. In one, hatching takes place very quickly and the young mature in about three days. In the other case eggs can lie dormant, surviving desiccation, and will hatch when more favorable circumstances prevail. The members of the class Macrodasyoidea are bisexual. Like the kinorhynchs, gastrotrichs are of little ecological or economic importance.   AC

Head
Jaw
Digestive gland
Trunk
Gonad
Bladder
Pedal gland
Gripping toe
Corona
Eye spot
Mastax
Corona retractor muscle
Stomach
Nephridiophore
Anus
Foot
1

Mouth cone
Head spines
Intestine
Tail spine
Mouth
2

Head
Neck
Trunk
Cilia
Papillae
Forked tail
3

important attribute for species dwelling in temporary pools of water. They can also serve as a dispersal phase as they can be blown about by the wind. When they eventually hatch only females develop from them. Survival over difficult times is also helped by the ability of many rotifers to tolerate water loss and to shrivel up into small balls and thus remain in a state of suspended animation. This ability is referred to as cryptobiosis or "hidden life".

LAMPSHELLS

Phylum: Brachiopoda
About 260 species in 69 genera and 2 classes.
Distribution: worldwide, marine.
Fossil record: very extensive, early Cambrian (about 600 million years ago) to recent with greatest profusion in the late Paleozoic (up to 225 million years ago).
Size: shell length from less than 0.04–1.5in (1mm to 4cm) plus stalk. Some extinct species reached 12in (30cm).

Features: stalked animals with bilaterally symmetrical shells in two halves (bivalve) comprising dorsal and central valves; body cavity a coelom; circulatory system open with contractile dorsal vessel; excretion via paired nephridia which also serve to release the reproductive cells; conspicuous loop-shaped ciliated respiratory and filter-feeding tentacle crown (lophophore).

Class: Articulata
Three living orders, 4 extinct. Valves locked at rear by a tooth and socket; lophophore has internal support; anus absent; includes genus *Terebratella*.

Class: Inarticulata
Two living orders, 3 extinct. Valves held together by muscles only; lophophore lacks an external skeleton; anus present; includes genus *Lingula*.

▶ **A covey of lampshells** (genus *Terebratulina*). It is just possible to make out the lophophore, with its many fine tentacles, inside the shells.

▼ **The precursor of a juvenile lampshell:** a larva of the genus *Lingula*. The large disk-like structure marks the development of the mantle and shells while the developing lophophore crown shows as the group of tentacles. As the shell gets heavier the lava sinks and takes up adult life without metamorphosis.

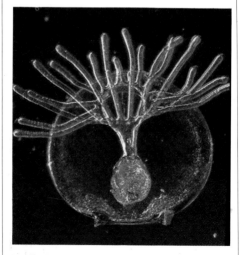

▶ **Lampshells:** ABOVE exterior view, attached to the substrate; BELOW the body plan.

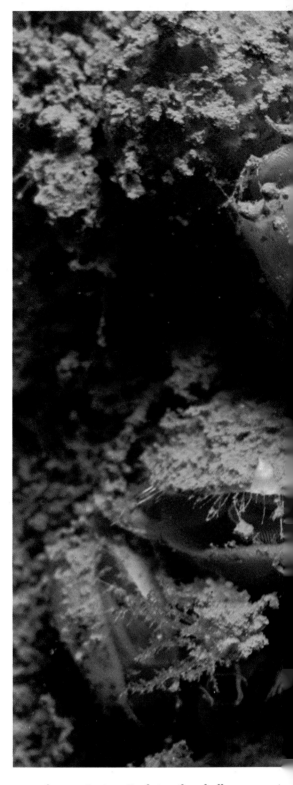

THE common name "lampshell" refers to the superficial resemblance between the shells of some of these animals and the Roman oil lamp. Lampshells are really animals of the past, for although they are present in modest numbers in various marine habitats the world over nowhere today do they dominate the seas of the world as they did in the late Paleozoic. Over 30,000 fossil species of lampshells have been identified between Cambrian and recent times.

Lampshells resemble bivalve mollusks (clams, mussels etc) and indeed were classified with them until the middle of the 19th century. Their bodies are ensheathed in a mantle and enclosed by two hinged valves (shells) which protect the soft animal within. The two lobes of the mantle secrete the shells and also enclose and protect the crown (lophophore). However, there are some startling differences between lampshells and mollusks which begin with the issues of symmetry and orientation.

Today lampshells are found living mainly on the continental shelves either attached to rocks or other shells (for example *Crania*) or some (for example *Lingula*) living in burrows in mud. Attachment may be direct or by a cord-like stalk. Some occur in very shallow water, even on shores. Few prosper at any depth. The distribution of these animals is sporadic but where they do occur they may exist in great numbers. The anatomy of lampshells is variable and quite complex. A salient feature is that the upper (dorsal) valve is smaller than the lower (ventral) one. In most forms the shells are both convex, and the apex of the lower one may be extended to give the effect of the spout of the Roman lamp that the common name alludes to. In some of the burrowing forms the shells are more flattened. The shells themselves may be variously decorated with spines, concentric growth lines fluted or ridged, and their colors vary from orange and red to yellow or gray.

The two valves are hinged along the rear line and the manner of their contact forms the basis for a division of the phylum into two classes: Inarticulata and Articulata. In the inarticulate lampshells the valves are linked together by muscles in such a way that they can open widely, but in the articulate lampshells they carry interlocking processes and these limit the extent of the gape. In inarticulate lampshells as many as five pairs of muscles control the movements of the valves while in articulates two or three sets of muscles are involved.

There are other differences between the two classes. In inarticulates the shells are formed from calcium phosphate and the gut terminates in a blind pouch, there being no anus. In articulates the shell is made of calcium carbonate and the gut is a "through" system with an anus. The stalk or pedicle by which most species are attached to the substrate is itself attached to the lower valve. The animals often position themselves so that the lower valve is uppermost, thus adding confusion to the ideas of orientation and symmetry. In a few species the pedicle has been completely lost and here the animals are directly attached to the substrate by the lower valve with the upper valve uppermost.

margin of the valves. They tend to be held close to the upper valve. Suspended food material is trapped in mucus and swept to the mouth via a special groove. Special currents carry away rejected particles. The mouth leads to an esophagus which in turn gives on to the stomach.

There is an open circulatory system with a primitive pumping heart situated above the stomach. Blood channels supply the digestive tract, tentacles of the lophophore, the gonads and the nephridia. There is no pigment and there are few blood cells. Circulation of abolished food is likely to be the main function of the blood system and oxygen transport seems to be the responsibility of the coelomic fluid.

Lampshells have one or two pairs of nephridia for excretion and these discharge through pores situated near the mouth. The nervous system is rather simple with a nerve ring around the esophagus and a smaller dorsal and larger ventral ganglion. From these ganglia nerves supply the lophophore, mantle lobes and the various muscles that control the valves.

There are very few hermaphrodite lampshells. There are generally up to four gonads per individual and the ripe gametes are discharged into the coelom and leave the body via the nephridiopores. The eggs are generally fertilized in the sea. In most cases the embryo develops into a free-swimming larva, but in a few species it may be brooded. The larval development and planktonic period vary considerably between species.

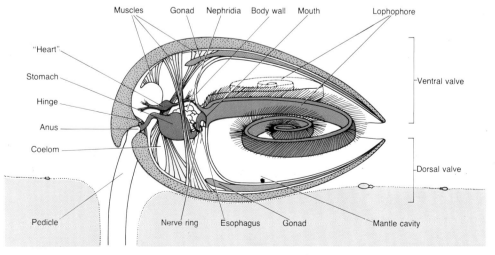

AC

Setae

Lower (ventral) valve

Upper (dorsal) valve

Pedicle

The body of the animal lies within the valves sheathed in mantle tissue which of course are extensions of the body wall. A true body cavity (coelom) lies between the body wall and the gut. The epidermis of the exposed inner mantle surface is hairy (ciliated). The lophophore is suspended in the mantle cavity and consists of a folded crown of hollow tentacles surrounding the mouth. It is supported by the upper valve of the shell and in some species there is actually a special skeletal structure to carry it. For feeding, the valves gape to the front allowing water to flow over the lophophore. The individual tentacles, which are hollow and ciliated on the outside, can reach to the front

Muscles   Gonad   Nephridia   Body wall   Mouth   Lophophore

"Heart"

Stomach

Hinge

Anus

Coelom

Ventral valve

Dorsal valve

Pedicle   Nerve ring   Esophagus   Gonad   Mantle cavity

MOSS ANIMALS

Phylum: Bryozoa (or Ectoprocta)
Moss animals or sea mats.
About 4,000 species in about 1,200 genera and 3 classes.
Distribution: worldwide, some freshwater but mainly marine; sedentary bottom-dwelling animals found in all oceans and seas at all depths but mainly between 165 and 655ft (50–200mm).
Fossil record: from Ordovician era (500–440 years ago) to recent times; about 15,000 species.
Size: individuals 0.01–0.06in (0.25–1.5mm), colonies 0.2–8in (0.5–20cm) wide.

Features: microscopic colonial animals attached to a substrate, permanently living inside a secreted tube or case (the zoecium); body cavity a coelom; mouth encircled by a hairy (ciliated) tentacle crown (lophophore); gut U-shaped; anus outside the lophophore; no nephridia or circulatory system.

Class: Gymnolaemata
Two orders.
Marine. Including genera *Alcyonidium*, *Memranipora*.

Class: Phylactolaemata
One order. Fresh water.

Class: Stenolaemata
Four living orders, three extinct. Marine. Including genus *Crisia*.

► **Colonial conditions:** BELOW individual zooids of bryozoans (genus *Flustrella*) protrude their ciliated lophophores from the colony surface.

▼ **Colonial structure:** an encrusting colony of bryozoans (genus *Electra*).

THE moss animals are an important group of sedentary aquatic invertebrates. They live attached to the substrate, either on rocks or empty shells, or on tree roots, weeds or other animals where they can find a hold. They are of interest because of their number and diversity. Some types, for example *Flustra*, may dominate conspicuous regions of the sea bed and have particular groups of organisms living alongside them. Others, such as *Zoobotryon*, are important because they can foul man-made structures, for example piers, pilings, buoys and ships' hulls.

Most moss animals are marine, inhabiting all depths of the sea from the shore downwards, in all parts of the world. A smaller group inhabit fresh water, where they are relatively inconspicuous. Moss animals are colonial animals. This means that many individuals termed zooids live together in a common mass. In a colony the first individual (formed from the larva, itself the result of sexual reproduction) is the founder. From the founder all other individuals in the colony are produced by asexual reproduction and thus share a common genetic constitution.

In many colonies of moss animals the individuals are all similar and all participate in feeding and sexual reproduction. Each individual is housed in a protective cup (zoecium), with walls reinforced with gelatinous horny or chalky secretions. The nature of these individual cup walls is conferred on the texture of the colony as a whole, so that some like *Alcyonidium* may feel soft and pliant and some like *Pentapora* feel hard, sharp and stiff. Equally the form of the colonies varies. Some genera, such as *Electra*, adopt a flat encrusting habit and therefore become dependent on currents flowing very close to the rocks to bring them their food, while others like *Myriapora* grow up from the substrate so as to exploit stronger currents flowing above the substrate.

Each feeding individual in a colony is constructed on a similar plan. There is no head as such. A crown of hollow tentacles, filled by the coelom, forms a food-collecting and respiratory surface. (This parallels the tentacles of lampshells and horseshoe worms.) This crown is termed the lophophore. At the center of the lophophore the mouth opens and leads in to a simple U-shaped gut with esophagus and stomach. The anus lies close to, but *outside* the lophophore—the ectoproct condition (compare the endoprocts, p48, where the anus too lies inside the ring of tentacles). (It should perhaps be added here that opinions vary widely on the relationship of the ectoprocts and endoprocts, but there are good reasons for considering them to be separate phyla.)

The lophophore can be withdrawn for protection by a set of muscles and it can be extended by other muscles which deform one wall of the cup. This is often the upper or frontal wall, which includes a flexible, frontal membrane. When special muscles pull the frontal membrane in, fluid pressure in the coelom increases and the lophophore is protracted. Many interesting evolutionary developments have been explored by moss animals to ensure that a flexible frontal membrane to the cup, or some other means of adjusting the cup volume, is retained while making sure that predators are deterred from gaining entry through what is potentially a weak spot. Clearly rigid protection and flexibility are not easily reconciled. It is the epidermis of the animal that secretes the horny or gelatinous cuticle and in the chalky ectoprocts this in turn is

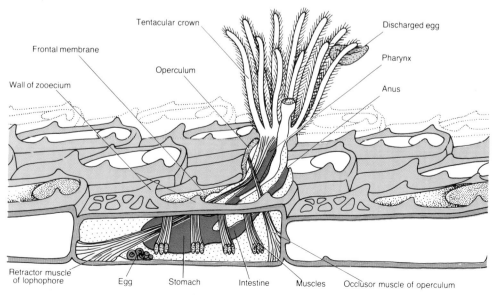

Tentacular crown
Frontal membrane
Operculum
Wall of zooecium
Discharged egg
Pharynx
Anus
Retractor muscle of lophophore
Egg
Stomach
Intestine
Muscles
Occlusor muscle of operculum

▶ **A foliose hydrozoan** attached to the rocky seabed. The skeletal system of rigid bones protecting the zooids helps this colony maintain its shape in the current. The lophophore can be discerned sieving the seawater.

▼ **The forms of some genera** of colonial moss animals. (1, 2) *Bugula*. (3) *Flustra*. (4) *Flustrellidra*. (5) *Myriapora*. (6) *Cupuladria*. (7) *Sertella*. (8) *Cellaria*. (9) *Pentapora*. (10) *Alcyonidium*.

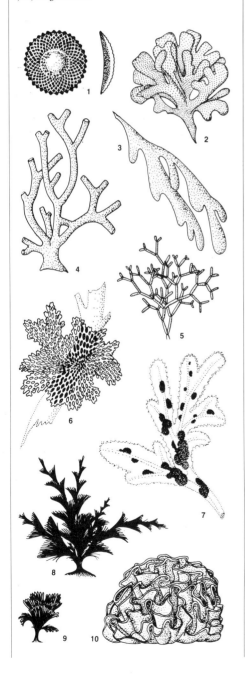

reinforced with calcium salts.

The lophophore tentacles are covered with waving hairs (cilia) to make efficient filter-feeding organs. When the lophophore is withdrawn the hole it goes in by may be covered over with a flap-like operculum or lid, present in some groups, but not in others.

Most animals are hermaphrodite with male and female organs developing from special regions of the inner lining of the body cavity. Ripe sperm and eggs are liberated into the coelom and find their way out of the body by separate pores. In many species the fertilized egg is brooded in a large chamber or ovicell, but in some a free-swimming larva, in many cases called a cyphanautes, spends a period in the plankton feeding on minute algae before settling, metamorphosing and founding a new colony.

The body contains no circulatory or excretory mechanisms; as it is of such small volume these roles are undertaken by diffusion.

In some of the more highly evolved moss animals (class Gymnolaemata, order Cheilstomata) there are several forms in the colonies. Here some zooids have given up the role of feeding and are supplied with food by others. Instead they function to defend the colony against small creeping predators like amphipod crustacea, and against clogging up with silt. There are two such forms: aviculariae and vibraculae. The first are so-called because they resemble minute birds' heads with snapping beak-like jaws. The cup and the operculum have been modified to act as minute jaws with delicate sense organs. They can seize hold of and detain small animals. The others have a large elongated operculum which resembles a bristle and which is vibrated back and forth to act as a current generator sweeping away silt.

AC

HORSESHOE WORMS

Phylum: Phoronida
About 11 species (more known as larvae only)
in 2 genera.
Distribution: marine, bottom-dwelling, mainly
from shallow temperate and tropical seas.
Fossil record: none.
Size: 0.2–8in (0.5–20cm).

Features: bilaterally symmetrical worm-like
animals living permanently in a secreted tube;
body cavity a coelom; mouth surrounded by a
ciliated tentacle crown (lophophore); gut
U-shaped; anus not enclosed by lophophore;
closed circulatory system containing red blood
corpuscles; paired nephridia also serve as ducts
for release of reproductive cells.

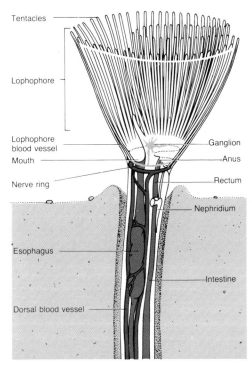

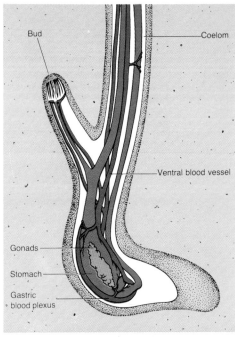

UNLIKE the moss animals, to which they are probably related, the horseshoe worms form a small and relatively insignificant phylum. Only some 11 species are known as adults and they are not of any ecological or commercial importance.

Horseshoe worms live attached to the substrate, in secreted tubes of chitin-like substance which quickly become decorated with fragments of shells and grains of sand. These animals are small and obscure and are normally discovered only by accident when for example items of substrate are being carefully examined. They live in shallow temperate and tropical habitats. None dwell in fresh water. Their known distribution in America, Australia, Europe and Japan is more likely a reflection of where marine biologists are active in discovering them rather than an accurate delineation of where they actually dwell. A number of larvae, technically known as actinotrochs, have been collected, the adults of which are unknown. It is certain that the development of marine biology will see an increase in the number of known species of horseshoe worms.

The horseshoe worm body is essentially worm-like. At the front is a well-developed horseshoe-shaped lophophore which in plan resembles two crescents, a smaller one set inside a larger one with their ends touching. The mouth lies between the two crescents. From these crescent-like foundations a number of ciliated tentacles arise to form the lophophore crown, and in some species the crescent tips may be rolled up as spirals thus increasing the extent of the lophophore. The beating of the cilia drives currents of water down the tentacles and out between their bases. Any small food particles are rolled up with secreted mucus and passed to the mouth.

The gut itself is U-shaped, a long esophagus leads back from the mouth to the stomach which is situated near the rear of the animal. From this the intestine leads back to the front and the rectum opens by the anus which is situated on a small protrusion (papilla) just outside the lophophore. The body cavity, a true coelom, houses paired excretory nephridia whose openings lie near the anus, and the gonads. Most species are hermaphrodite, and the sex cells are liberated into the coelom and find their way to the exterior via the excretory ducts.

The body wall consists of an outer tissue (epithelium) covering a thin sheath of circular muscles, which when they contract make the animal long and thin. Inside the circular muscles is a thick layer of longitudinal muscle which is responsible for contracting and shortening the body. The muscles enable the animal to move inside the tube and they are controlled by a simple nervous system. There is a nerve ring at the front set in the epidermis at the base of the lophophore. From this nerves arise and innervate the lophophore tentacles and the muscles of the body wall.

One interesting feature of the horseshoe worms is their clearly defined blood system with red pigment (hemoglobin) borne in corpuscles. Blood is carried forward by a dorsal vessel which lies between the limbs of the gut and back to the posterior by a ventral vessel. Small branches penetrate each tentacle of the lophophore. The blood is moved around the system by contractions that sweep along the dorsal and ventral vessels.

▲ **Water fans:** a colony of horseshoe worms feeding. Hemoglobin is visible in the blood vessels of some of the individuals.

◄ **The internal structure** of a horseshoe worm (genus *Phoronis*).

One species, *Phoronis ovalis*, is known to reproduce asexually and to establish large aggregations by budding. In sexual reproduction the eggs are generally fertilized in the sea. Typically an actinotroch larva develops which lives in the plankton for several weeks before undergoing rapid change to settle on the sea bed, form a tube and take up the adult mode of life.

One aspect of interest which zoologists attach to the horseshoe worms is their evolutionary position in general and their relationship in particular with the phyla of moss animals and lampshells. In the course of development embryos of species in all these phyla show a form of development where the body becomes divided into three sections, each with its own region of body cavity (coelom). As the animals develop the front section becomes progressively reduced and all but disappears. The middle region with its own coelom forms the lophophore and the rear section forms the bulk of the body of the adult. Zoologists have described these animals as oligomerous, that is they have few sections or segments. They are thus quite different from those groups like the annelid worms and the arthropod phyla where there are many segments to the body, ie metamerous. The general pattern of development in these groups is inclined toward that of the protostome groups, for example the annelid worms, but occasional features are inconsistent with this suggesting that in evolutionary terms they may occupy a position among the coelomate groups intermediate between the annelids and the arthropod phyla on the one hand, and the echinoderms on the other.

AC

Acoelomate

Flatworms
Phylum: Platyhelminthes

Ribbon worms
Phylum: Nemertea

Pseudocoelemate

Roundworms
Phylum: Nematoda

Spiny-headed worms
Phylum: Acanthocephala

Horsehair worms
Phylum: Nematomorpha

Coelomate – segmented

Segmented worms
Phylum: Annelida

Coelomate – not segmented

Echiurans
Phylum: Echiura

Priapulans
Phylum: Priapula

Sipunculans
Phylum: Sipuncula

Beard worms
Phylum: Poganophora

Arrow worms
Phylum: Chaetognatha

Acorn worms
Phylum: Hemichordata

WHAT is a worm? Everyone thinks they know how to deal with this question, but in reality it is a difficult one to answer. The standard reply will refer to earthworms—after all they are the most familiar invertebrate animals as they occur in our gardens and pasture lands. The sea angler will think of marine worms too, valuable as bait, like lugworms and ragworms. Others will remember the leeches, renowned as parasites, but also living as predators. Lugworms, ragworms, earthworms and leeches are all examples of one of the best known groups of worms, the segmented or annelid worms. This group contains a great variety of forms, marine and freshwater, terrestrial and parasitic, but they are not the only worms in the animal kingdom. The typical worm is a long, thin animal with a variably developed head and tail. It is therefore bilaterally symmetrical, unlike many of the creatures described in the previous sections on sedentary and free-swimming invertebrates which were asymmetrical or radially symmetrical.

Since worms have evolved at many evolutionary levels in the animal kingdom, there are many variations on the worm "theme." It is easiest to understand the different levels to which worms have evolved by looking at their structure in terms of number of body layers, origins and types of body cavities and their disposition, as well as observing whether or not the animals are divided into a number of segments. The simplest structure is found in flatworms and ribbon worms, where there is no internal body space. In roundworms and spiny-headed worms a body cavity is present between the gut and body wall, but this space is not lined on its inner edge by mesoderm (the middle tissue block) and is hence termed a pseudocoel. In the earthworms, ragworms, leeches and various other phyla the body cavity is known as a true coelom, since it is formed within the mesoderm and hence banded by it on both sides. Of those worm-like animals

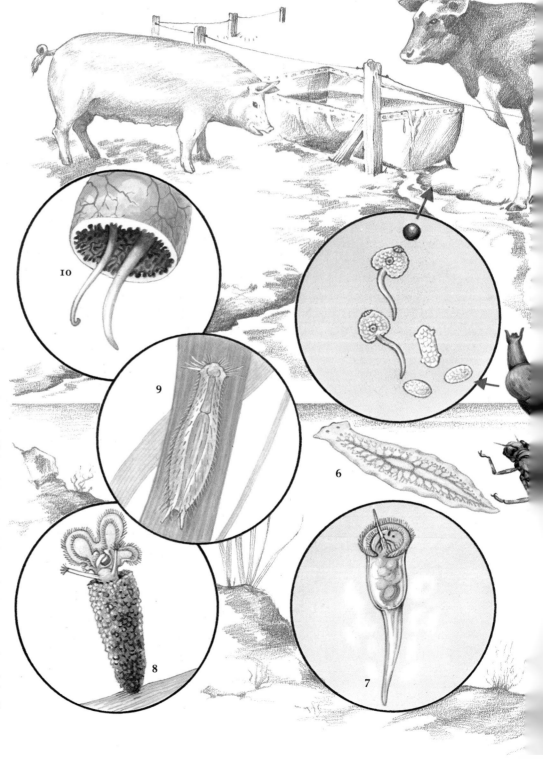

VERTEBRATES

▼ **Worm-like parasites** of farmland and some other minute invertebrates. (1) *Fasciola hepatica*, the common liver fluke, a commercially significant parasite of sheep and cattle in Europe. The adults live in the liver. Eggs pass out of the primary host in feces and hatch into the first larval stage, a miracidium, which swims by cilia in the surface film of water on vegetation. The miracidium detects a secondary host, the fresh water snail *Lymnaea truncatula*, and enters it, forming a cyst in the digestive gland. From this a third larval stage, the rediae, develops, followed by the cercariae which leave the snail. These may encyst prior to reinfection of the primary host, thus completing the life cycle. The liver fluke flourishes in damp pastures where the snails exist freely. Adult size up to about 1.2in (3cm). (2) A species of *Brachionus*, an example of a rotifer protected by a firm outer structure called a lorica. (3) *Stephanoceros fimbriatus*, a sessile rotifer protected by a gelatinous case. (4) *Dugesia subtentaculata*, a free-living flatworm (planarian). (5) A species of *Gordius*, a parasitic worm (2in, 5cm). (6) *Dendrocoelum lacteum*, a free-living flatworm (planarian). Like *Dugesia* it glides through the environment by means of its cilia. (7) *Conochilus hippocrepis*, an example of a free-living rotifer protected by a gelatinous case. (8) *Floscularia ringens*, a rotifer that builds itself a protective tube out of pellets (microscopic). (9) A species of *Chaetonotus*, an example of a gastrotrich. (10) *Ascaris lumbricoides suilla*, a round worm parasite of the gut of the domestic pig, widely distributed (length about 1in (2.5cm).

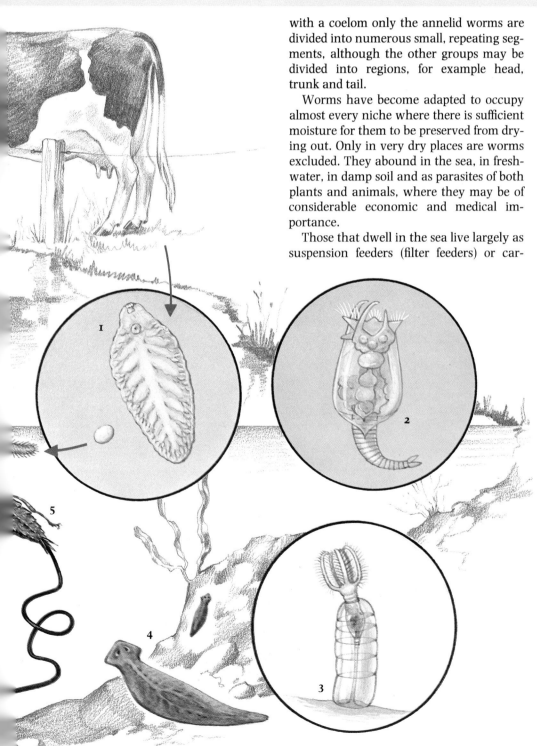

with a coelom only the annelid worms are divided into numerous small, repeating segments, although the other groups may be divided into regions, for example head, trunk and tail.

Worms have become adapted to occupy almost every niche where there is sufficient moisture for them to be preserved from drying out. Only in very dry places are worms excluded. They abound in the sea, in freshwater, in damp soil and as parasites of both plants and animals, where they may be of considerable economic and medical importance.

Those that dwell in the sea live largely as suspension feeders (filter feeders) or carnivores. The suspension feeders are equipped with filters and either depend on currents to sweep food to them, like the polychaetes *Sabella* and *Serpula* or horseshoe worms such as *Phoronis*, or, like the polychaete *Chaetopterus*, use modified limbs to generate a current to pump water through their filters. Carnivores need to subdue and to capture their prey. Ribbon worms have a long proboscis which is sometimes armed with stylets and glands which can be coiled around the victim. The annelids include forms which have powerful jaws like *Nereis* and *Marphysa*. By complete contrast the arrow worms are planktonic and not bottom dwellers. They too are efficient predators, their mouths being armed with seizing teeth which can grip small drifting organisms and readily engulf them. Feeding on detritus and swallowing mud are common ways of life for bottom dwellers in the sea and many annelids as well as the sipunculans and echiurans have adopted such strategies, often simply swallowing the nutrient-rich deposits through which they burrow. The acorn worms too live in muddy deposits and use the gills sited in their pharynx to filter water currents for food. Possibly the most enigmatic life styles of the free-living worms are those of the beard worms which live in tubes buried in the mud of the ocean bed. These animals lack a gut but have a crown of tentacles. They appear to absorb their nutrient requirements from the surrounding sea water.

Many worms have evolved to a parasitic mode of life. Without doubt the parasitic flatworms, flukes and tapeworms, have achieved great success deploying complex life cycles to ensure reinfection of their hosts. The spiny-headed worms are a small, exclusively parasitic group that have no gut, absorbing their nutrients directly from their hosts. Many roundworms feed as external or internal parasites on the bodies of plants and animals using special stylets and teeth to pierce and suck.   AC

FLATWORMS AND RIBBON WORMS

Phyla: Platyhelminthes, Nemertea

Flatworms
Phylum: Platyhelminthes.
About 5,600 species in 3 classes.
Distribution: worldwide; free-living aquatic;
parasitic in invertebrates and vertebrates.
Fossil record: earliest in Mesozoic era 225–65
million years ago.
Size: microscopic (1μm) to 13ft (4m), mostly
0.004–0.4in (0.01–1cm).

Features: bilaterally symmetrical, usually
flattened, the first animals to have evolved front
and rear ends and upper and lower surfaces;
triploblastic, having a third layer of cells
(mesoderm) which, by itself or in combination
with ectoderm or endoderm, gives rise to
organs or organ systems; without skeletal
material, true segmentation, body cavity
(acoelomate) or blood-vascular system; defined
head with a concentration of sense organs and
a central nervous system; excretory vessels
originate in ciliary flame cells; gut may be
absent, rudimentary or highly developed but
there is no anus; sexual and asexual
reproduction, male and female organs often in
the same animal, often hermaphrodite;
development sometimes direct, sometimes
accompanied by metamorphosis.

Ribbon worms
Phylum: Nemertea.
Ribbon or proboscis worms.
About 750 species in 2 classes.
Distribution: worldwide; free-living, mostly
marine but some in freshwater and damp soil.
Fossil record: as for platyhelminthes.
Size: from 0.08in–85ft (0.2cm to 26m).

Features: bilaterally symmetrical, unsegmented
worms with flattened elongated body; with a
ciliated ectoderm; without a distinct coelom
(pseudocoelomate); similar to flatworms but
more highly organized; eversible proboscis in a
sheath; gut with an anterior mouth, distinct
lateral diverticula and anus at rear; there is a
blood-vascular system and excretory vessels
with ciliary flame cells; reproduce mostly by
sexual means, occasionally asexual (by
fragmentation), sexes separate; development in
some direct, in others accompanied by
metamorphosis.

▶ **Sliding smoothly** OPPOSITE through a bed of
sea squirts, a Banded planarian (*Prostheceraeus
vittatus*). Most free-living planarian flatworms
such as this marine species avoid strong light
and hide under stones or vegetation.

▶ **Body plan of a planarian flatworm** (genus
Planaria), showing digestive, excretory,
reproductive and nervous systems.

FLATWORMS were the first animals to
evolve distinct front and rear ends. They
are also the simplest of the animals with a
three-layered (triploblastic) arrangement of
the body cells. The most remarkable groups
are the parasitic flukes and tapeworms, but
many species are free living.

The turbellarians (class Turbellaria) are a
widely distributed group of mostly free-
living flatworms with a ciliated epidermis.
They usually possess a gut but have no
anus, and are classified by the shape of the
digestive tract. Best known are the
planarians, which may be black, gray or
brightly colored, are aquatic in fresh or sea
water, and feed on protozoans and small
crustaceans, snails and other worms. *Convoluta* has no gut and depends on symbiotic
algae for its food. Some members of the order
Rhabdocoela (eg *Temnocephala* species) live
attached to crustaceans and to other invertebrates and feed on free-living organisms,
but most have become parasitic, such as
Fecampia species, which live inside crabs and
other crustaceans.

Some planarians regenerate complete
worms from any piece. The more sophisticated parasitic flatworms, however, are
unable to replace lost parts. In general the
lower the degree of organization of an
animal, the greater its ability to replace lost
parts. Even a small piece, cut from such an
animal usually retains its original polarity:
a regenerated head grows out of the cut end
of the piece which faced the front end in the
whole animal. The capacity for regeneration
decreases from front to rear: pieces from
forward regions regenerate faster and form
bigger and more normal heads than pieces
from further back. In some planarians only
pieces from near the front are able to form

a head; those further back effect repair but
do not regenerate a head.

The head of a planarian is dominant over
the rest of the body. Grafts of head pieces
reorganize the adjacent tissues into a whole
worm in relation to themselves. Grafts from
tail regions, on the other hand, are generally
absorbed. However, the dominance of the
head over the rest of the body is limited by
distance, for example, if the animal grows
to a sufficient length. This is what happens
when planarians reproduce asexually: the
rear part starts to act as if it were "physiolo-
gically isolated" and then finally constricts
off as a separate animal.

True parasitism is universal in the two
other classes, the Trematoda and Cestoda.
The majority of monogenean trematodes
are external parasites living on the outer
surface of a larger animal, many on the gills
of fishes. A few inhabit various internal
organs and are true internal parasites, such
as *Polystoma* in the urinary bladder of a frog
and *Aspidogaster* in the pericardial cavity of
a mussel species. The digenean trematodes,
including the flukes, are all internal parasites. The adults inhabit in most cases the
gut, liver or lungs of a vertebrate animal,
swallowing and absorbing the digested food,
blood or various secretions of their host.
Among these are *Fasciola* species in the liver
of sheep and cattle, *Schistosoma* in the blood
of man and cattle, and *Paragonimus* in the
lungs of man. The internal parasites are
parasitic throughout the greater part of
their life. After an initial short period as a
free-living ciliated larva known as a miracidium, the young enters a state of parasitism as a sporocyst or redia in a second host,
and after a second free interval as a tadpole-
shaped cercaria, may enter the body of a
third host to become encysted. The second
host is often a mollusk, and the cercaria may
complete its life cycle by becoming encysted
in the same animal or a fish.

The cestodes are the most modified for a
parasitic existence, as they remain internal
parasites throughout life. They invariably
inhabit the gut of a vertebrate. The intermediate host is frequently also a vertebrate—
commonly the prey of the final host. As an
adult, *Taenia crassicollis* is parasitic in the
intestine of the cat; the cysticercus larval
stage occurs in the livers of rats and mice.
The adult tapeworm *Echinococcus granulosus*
inhabits the gut of dogs and foxes; its
hydatid cyst larval stage may be found in
the liver or lungs of almost any mammal,
but especially sheep and occasionally man.

Flatworms are bilaterally symmetrical
animals without true segmentation. The

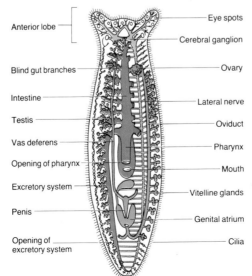

Anterior lobe — Eye spots
— Cerebral ganglion
Blind gut branches — Ovary
Intestine —
Testis — Lateral nerve
Vas deferens — Oviduct
Opening of pharynx — Pharynx
— Mouth
Excretory system — Vitelline glands
Penis —
— Genital atrium
Opening of — Cilia
excretory system

Flatworms and Ribbon Worms

Flatworms
Phylum Platyhelminthes

Free-living flatworms
Class Turbellaria
About 1,600 species.

Order Acoela, eg *Convoluta*.
Order Rhabdocoela, eg *Dalyellia*,
Fecampia, *Mesostoma*, *Temnocephala*.
Order Tricladida (**planarians**),
eg *Dendrocoelum*, *Planaria*, *Polycelis*,
Procerodes, *Rhynchodemus*.
Order Polycladida (**planarians**),
eg *Planocera*, *Thysanozoon*.

Parasitic flatworms
Class Trematoda
About 2,400 species.

Subclass Monogenea
—**monogeneans**
Order Monopisthocotylea,
eg *Gyrodactylus*.
Order Polyopisthocotylea,
eg *Polystoma*.

Subclass Aspidogastrea
Eg *Aspidogaster*

Subclass Digenea—**flukes,
digeneans**
Order Strigeatoidea, eg *Alaria*,
Schistosoma.
Order Echinostomida, eg *Echinostoma*,
Fasciola.
Order Opisthorchiida, eg *Heterophyes*,
Opisthorchis.
Order Plagiorchiida, eg *Paragonimus*,
Plagiorchis.

Tapeworms
Class Cestoda
About 1,600 species.

Subclass Cestodaria
Order Amphilinidea, eg *Amphilina*.
Order Gyrocotylidea, eg *Gyrocotyle*.

Subclass Eucestoda
Order Tetraphyllidea,
eg *Phyllobothrium*.
Order Protocephala, eg *Proteocephalus*.
Order Trypanorhyncha,
eg *Tetrarhynchus*.
Order Pseudophyllidea,
eg *Dibothriocephalus*, *Ligula*.
Order Cyclophyllidea, eg *Echinococcus*,
Taenia.

Ribbon worms
Phylum Nemertea

Class Anopla
Order Palaeonemertini, eg *Carinella*.
Order Heteronemertini,
eg *Baseodiscus*, *Lineus*.

Class Enopla
Order Metanemertini,
eg *Drepanophorus*, *Prostoma*.

shape is leaf-like or ribboned in the planarians, or cylindrical in some rhabdocoels. While a distinct head is rarely developed, there is often a difference marking the anterior end, eg the presence of eyes, a pair of short tentacles, a slight constriction to form an anterior lobe. The mouth is on the underside, mostly toward the front. In some polyclads there is a small ventral sucker on the underside, and in some rhabdocoels there is an adhesive organ at the front and rear. In *Temnocephala* species a row of tentacles is often present. The trematodes closely related to the Turbellaria in internal organization resemble them in external form, with futher modifications to accommodate the parasitic existence. They are generally leaf-like with a thicker, more solid body. Suckers on the underside fix the parasite to its host. Usually there is a set at the front, or a single sucker surrounding the mouth and a rear set, or a single large rear sucker. Among trematodes, monogeneans often have more numerous suckers. Cestodes are ribbon-like, the anterior end is, in most cases, attached to the host by means of suckers and hooks placed on a rounded head (scolex). The hooks are borne on a retractile process, the rostellum. In the order Pseudophyllidea a pair of grooves takes the place of suckers and there are no hooks. In many cestodes parasitic in fishes the head bears four prominent flaps, the bothridia. In *Tetrarhynchus* species there are four long narrow rostella covered with hooklets. The cestodes are mouthless, and nothing distinguishes upper and lower surfaces. The body or strobila, which is narrower at the front, is made up of a series of segment-like proglottides which become larger toward the rear.

The outer surface of the body wall of parasitic flatworms is differently modified in the three classes and the underlying layers of muscle are also differently arranged. A characteristic of flatworms is the mesenchyme, a form of connective tissue filling the spaces between the organs. There is an alimentary canal in the turbellarians (except the Acoela) and in the trematodes, which also have a muscular pharynx and an intestine. There is no gut in the cestodes, which take in nutrients through the surface of the body wall. Flatworms have a bilateral nervous system involving nerve fibers and nerve cells. The degree of development of the brain varies in the different groups, being greatest in the polyclads and some monogenean trematodes: there is a grouping of nerve cells at the front end into paired cerebral ganglia, especially in the free-living forms. Sense organs in adult free-living turbellarians include light, chemo-sensory and chemo-tactic receptors, and these may also occur in the free-living stages of trematodes and cestodes. An excretory system exists in nearly all flatworms except the Acoela. There is usually a main canal running down either side of the body, with openings to the exterior. Opening into the twin canals are small ciliated branches that finally end in an organ known as a flame cell. There is no circulatory-vascular system.

Both male and female reproductive organs occur in the one animal, which is usually hermaphrodite, one of the exceptions being the trematode genus *Schistosoma*. The reproductive organs are most complex in the parasitic forms. The testes are often numerous, their united ducts leading into a muscular-walled penis resting in the genital opening. The female reproductive organs comprise the ovaries, which supply the ova, and the vitellarium, which supplies the ova with yolk and a shell. The ovaries discharge their ova into an oviduct

▲ **Gliding through the leaf litter,** a free-living flatworm on the forest floor in Trinidad. Although mostly thought of as either aquatic or parasitic, flatworms are quite common sights in damp tropical forests.

▶ **Banded ribbon worm on the seabed** (genus *Tubulanus*). The body of many ribbon worms is long and cylindrical and often brightly colored. Most species are carnivorous, usually capturing living invertebrate prey.

▼ **Body plan of a ribbon worm,** showing the proboscis lying outside the body.

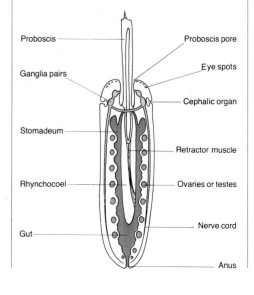

Proboscis

Ganglia pairs

Stomadeum

Rhynchocoel

Gut

Proboscis pore

Eye spots

Cephalic organ

Retractor muscle

Ovaries or testes

Nerve cord

Anus

Memory and Learning in Planarians

Planarians can learn and have a memory. They acquire a conditioned reflex in response to a bright light followed by an electric shock. If animals trained in this way are cut into two, both halves regenerate into a whole organism which retains its acquired learning, indeed both "heads" and "tails" show as much retention as uncut animals.

Similar results are found in planarians taught to find their way through a simple maze. Further experiments indicate that sexually mature worms fed on trained animals learn quicker than "control" animals kept on a normal diet. Two-headed planarians, produced by surgery, learn more quickly than others, while animals whose brain is removed are incapable of learning.

The memory of the animals cannot therefore be located in its brain and nervous system alone; it must be represented by chemical changes in cells throughout the body. It has been suggested that in the head of the planarian, memory is retained by neuron circuitry, whereas in the rest of the body it is retained in the form of a chemical imprint. This has been confirmed by an experiment in which trained worms were cut in two and made to regenerate in a liquid containing a chemical "memory eraser" known as ribonucleic acid ASE. The "heads" were not affected by it, but the "tails" forgot all they had learnt. Planarians show, beyond reasonable doubt, inheritance of acquired learning in animals that reproduce asexually.

which later forms a receptacle where fertilization occurs. The oviduct next receives the vitalline ducts which lead into the genital atrium. The location and arrangement of the genital opening in relation to the exterior are such to prevent self-fertilization and ensure cross-fertilization. Development in rhabdocoels and monogenean trematodes is direct. In digenean trematodes, cestodes and some planarians a metamorphosis occurs. Asexual reproduction occurs commonly in the turbellarians by a process of budding. Other planarians may fragment into a number of cysts each of which develops into a new individual.

The **ribbon worms** (phylum Nemertea) are almost entirely non-parasitic, marine worms, often highly colored, with only a few forms living on land or in freshwater, and one group as external parasites on other invertebrates. They are commonly looked upon as most closely related to the turbellarian flatworms, but are more highly organized.

Ribbon worms are often found burrowing in sand and mud on the shore or in cracks and crevices in the rocks. Some are able to swim by means of undulating movements of the body. Nearly all are carnivorous, either capturing living prey, mostly small intertebrates, or feed on dead fragments. The body is nearly always long, narrow, cylindrical or flattened, unsegmented and without appendages. The entire surface is covered with cilia and with gland cells secreting mucus which may form a sheath or tube for the creature. Beneath the ectoderm cell layer are two or three layers of muscle. There is no true coelom or body cavity, the space between organs being filled with mesenchyme. The proboscis, the most characteristic organ, lies in a cavity (rhynchocoel) formed by the muscular wall of the proboscis sheath. The proboscis may be everted for feeding. In representatives of the order Metanermertini (eg *Drepanophorus* species) there are stylets on the proboscis, providing formidable weapons. In other nemerteans (eg *Baseodiscus*) there are no stylets but the prehensile proboscis can be coiled round its prey and conveyed to the mouth underneath the front end. The gut is a straight tube from mouth to anus, but various regions are recognizable in some species. There is a blood-vascular system, the blood being generally colorless, with corpuscles. The excretory system resembles that of ribbon worms, but the nervous system is more highly developed. The brain is composed of two pairs of ganglia, above and below, just behind or in front of the mouth. Certain ganglia are probably related to special sense organs on the anterior end of the worm, and most ribbon worms have eyes. From the brain a pair of longitudinal nerve cords runs back down the body and there is a nerve net the complexity of which varies in the three orders. In most species the sexes are separate. The ovaries and testes are situated at intervals between the intestinal ceca and each opens by a short duct to the surface. Most ribbon worms develop directly, but some have a pelagic larva stage (pilidium), ending with a remarkable metamorphosis—the young adult develops inside the larva, from which it emerges.

Although very common, ribbon worms generally are of little economic or ecological importance.   GD

Pathogenic Parasites

Life cycles and medical significance of digenean flukes

Few animals affect humans so adversely as flukes of the subclass Digenea. They include creatures that cause one of the two most prevalent diseases of humans, schistosomiasis or bilharzia, and others that cause serious losses of livestock. These digenean flukes can be divided into blood flukes (eg *Schistosoma* species), lung flukes (eg *Paragonimus westermani*), liver flukes (eg *Fasciola* and *Opisthorchis* species), and intestinal flukes (eg *Fasciolopsis buskii*, *Heterophyes heterophyes*). Liver flukes can cause serious economic loss in sheep and cattle industries.

Other groups of parasitic flatworms contain representatives that, directly or indirectly, are harmful to people. Some monogenean parasites of fish often cause serious losses in fish-farming stocks kept in overcrowded conditions. Human cestode disease caused by adults of the tapeworm species *Taenia solium*, *T. saginata* and *Diphyllobothrum latum* is relatively nonpathogenic, but hydatid disease produced by the hydatid cysts of *Echinococcus granulosus*, cysticercosis caused by the cysticerus larval stage of *T. solium*, and sparganosis caused by the plerocercoid larvae of the genus *Spirometra* can be pathogenic.

The Common liver fluke (*Fasciola hepatica*) of sheep and cattle, almost worldwide in its distribution, is replaced by *F. gigantica* in parts of Africa and the Far East. Young flukes live in the liver tissue feeding mainly on blood and cells, while the adults live in the bile ducts. The flukes are hermaphrodite, with male and female organs in the same worm. Large numbers of eggs are produced and pass out in the feces onto the pasture. After a variable period, depending on the temperature, a miracidium hatches out. Moving on its cilia, this first-stage larva seeks out and penetrates the appropriate snail host, which in Europe and the USA is the Dwarf pond snail (*Lymnaea truncatula*), an inhabitant of temporary pools, ditches and wet meadows. The miracidium develops in its secondary host into a sporocyst which produces two or three generations of multiple rediae. Each redia produces free-living cercariae, the number being determined by temperature. From one miracidium therefore many thousands of infective cercariae are produced. The tadpole-shaped cercariae leave the snail and may encyst on grass as metacercariae until eaten by a suitable primary host, which may be a sheep, cow, donkey, horse, camel, etc and may also be a human eating, for example, infected watercress.

Spectacular losses due to acute fluke infes-

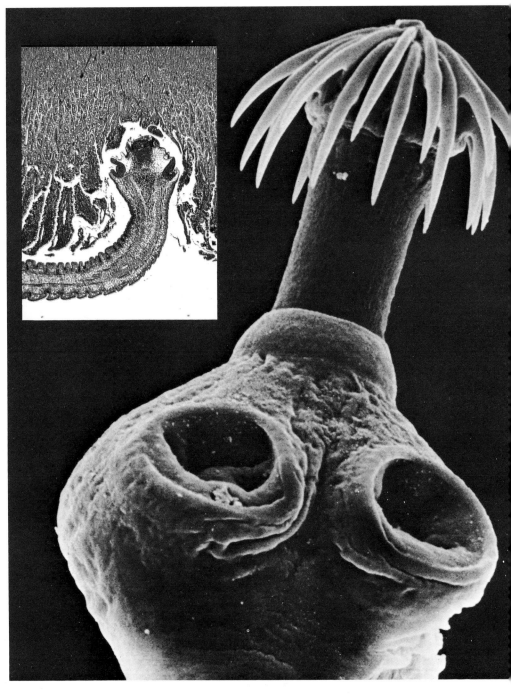

tation may occur with sheep but are rarer in cattle. The annual economic losses from fascioliasis in cattle and sheep in the UK for example are estimated to be approximately £10m and £50m respectively. Annual total losses in Holland may be 16m guilders, and in the USA $7m and in Hungary equivalent to $20m. It is possible in some countries to predict or forecast outbreaks of fascioliasis by using a meteorological system relying on monthly rainfall, evaporation and temperature data. In some countries chemical control of the snail host is possible, but in others reliance is placed on the periodic and strategic dosing of infected animals with such drugs as rafoxanide, oxyclozanide and

nitroxynil. A new breakthrough in treatment is triclabendazole which promises to be effective against early immature and mature *F. hepatica* in sheep and cattle.

Schistosomiasis or bilharzia affects the health of over 250 million people in 74 developing countries and it is estimated that another 600 million people are at risk. There are three principal schistosome species affecting humans. *Schistosoma haematobium* causes disease of the bladder and reproductive organs in various parts of Africa, especially Egypt, and the Middle East; *S. mansoni*, affecting mainly the liver and intestines, occurs in many parts of Africa, the Middle East, Brazil, Venezuela,

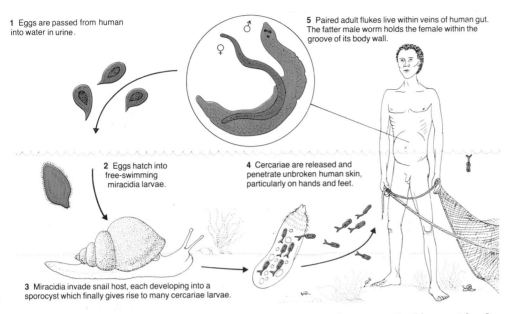

1 Eggs are passed from human into water in urine.

5 Paired adult flukes live within veins of human gut. The fatter male worm holds the female within the groove of its body wall.

2 Eggs hatch into free-swimming miracidia larvae.

4 Cercariae are released and penetrate unbroken human skin, particularly on hands and feet.

3 Miracidia invade snail host, each developing into a sporocyst which finally gives rise to many cercariae larvae.

▲ **Life cycle of the human blood fluke**
Schistosoma haemotobium which causes urinary or vesical schistosomiasis through damage to the bladder wall as its eggs bore through to get out of man. Water snails are the host for the larval stage.

◄ **Surreal palm tree with alien eyes**—the hooked, piston-like armory (rostellum) of the tapeworm *Acanthrocirrus retrirostris* taken in its larval form from a barnacle. Below the shaft on the head is a ring of four suckers, two of which are visible here. These are powered by flexible muscles and attach the tapeworm to the host tissue. The INSET shows the head of a dog tapeworm (*Taenia crassiceps*) embedded in the host's gut wall.

Surinam and some Caribbean islands; *S. japonicum*, also affecting the liver and intestines and sometimes the brain, occurs in China, Japan, the Philippines and parts of Southeast Asia.

Schistosomes live in the blood vessels feeding on the cells and plasma. They do little damage themselves but are prolific egg-layers. The eggs pass through the veins into the surrounding tissues of the bladder, intestines or liver, causing severe damage. The long cylindrical adult females live permanently held in a ventral groove of the more muscular males, laying eggs which eventually pass out in the host's urine or feces. Adult schistosomes possess adaptations which enable them to evade the immunological defenses of the host and they

may live and reproduce in one person for up to 30 years. Should the eggs reach fresh water they hatch, releasing a free-swimming miracidium which actively seeks out and penetrates the appropriate intermediate freshwater snail host—a pulmonate (eg *Bulinus*, *Physopsis*) or, for *S. japonicum*, a prosobranch snail. The parasite multiplies asexually inside the snail and after a period of between 24 and 40 days has produced a large number of infective cercariae. These escape from the snail and need to burrow through human skin or into another suitable host for the parasite to develop into either a male or female worm.

The geographical range of each species of schistosome is confined to the distribution of suitable snail hosts. The most important factor affecting the spread of schistosomiasis is the implementation of water-resource development projects in developing countries—primarily the construction of hydroelectric dams and also irrigation systems. The Aswan High Dam in Egypt, and Lake Volta in Ghana have already aggravated and increased the spread of a disease already endemic in these countries. The debilitating disease schistosomiasis and malaria are the two most prevalent diseases of the world and have received much attention from the World Health Organization (WHO), which stimulates research into the basic problems related to the spread and control of these important diseases. The WHO claims that schistosomiasis could be eliminated by a combination of clean-water projects and the intensive use of drugs such as metrifonate and a new drug (praziquantel) against the appropriate schistosomes. Some experts believe a vaccine is needed too, but as yet no such vaccine is available, although one may appear in the next few years as a result of applying new technology using monoclonal antibodies.

GD

▼ **Life cycle of the liver fluke** *Fasciola hepatica* in sheep. This parasite also infects cattle.

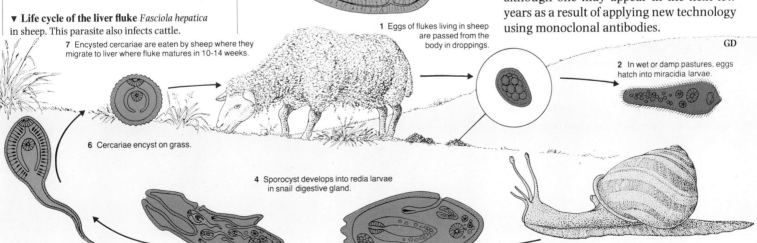

7 Encysted cercariae are eaten by sheep where they migrate to liver where fluke matures in 10-14 weeks.

1 Eggs of flukes living in sheep are passed from the body in droppings.

2 In wet or damp pastures, eggs hatch into miracidia larvae.

6 Cercariae encyst on grass.

4 Sporocyst develops into redia larvae in snail digestive gland.

5 Redia develop into cercaria larva which leave snail by pulmonary aperture.

3 Free-swimming miracidium penetrates snail *Lymnaea truncatula* where it developes into a sporocyst.

Phyla: Nematoda, Acanthocephala, Nematomorpha

Roundworms

Phylum: Nematoda

Roundworms, eelworms, threadworms or nematodes

About 12,000 named species, but many may remain to be discovered.

Distribution: worldwide; mostly free living in damp soil, freshwater, marine; some parasites of plants and animals.

Fossil record: sparse; around 10 species recognized in insect hosts from Eocene and Oligocene (54–26 million years ago).

Size: from below 0.02in to about 3.3ft (0.5mm–1m) but mostly between 0.04 and 0.08in (1–2mm).

Features: round unsegmented worms sheathed in resistant external cuticle; body tapers to head and tail; cavity between gut and body wall a pseudocoel; mouth and anus present; lack circular muscles, cilia, excretory flame cells and a circulatory system; nervous system quite well developed; sexes separate.

Class: Aphasmida
About 14 orders; includes genus *Trichinella*.
Class: Phasmida
About 6 orders; includes genus *Ascaris*, *Loa*, *Mermis*, *Onchocera*, *Wucheria*.

Spiny-headed worms

Spiny-headed or thorny-headed worms

Phylum: Acanthocephala

About 700 species in 3 classes.

Distribution: widespread as gut parasites of terrestrial, freshwater and marine vertebrates; arthropods are intermediate hosts.

Fossil record: only 1 species from Cambrian (600–500 million years ago).

Size: about 0.04in to 3.3ft (1mm–1m), but mainly below 0.8in (20mm).

Features: worms lacking a mouth, gut and anus; cavity a pseudocoel; peculiar protrusible proboscis ("spiny head") armed with curved hooks; sexes separate.

Class: Arahiacanthocephala
Two orders; includes genus *Gigantorhynchus*.
Class: Palaeacanthocephala
Two orders; includes genus *Polymorphus*.
Class: Eoacanthocephala
Two orders; includes genus *Pallisentis*.

Horsehair worms

Horsehair, gordian or threadworms

Phylum: Nematomorpha

About 80 species.

Distribution: worldwide; adults free living and primarily aquatic in freshwater; juveniles parasitic in arthropods; *Nectonema* marine.

Fossil record: none.

Size: 2–39in (5–100cm) long.

Features: unsegmented worms with thick external cuticle; front end slender; body cavity a pseudocoel; through gut with mouth and anus present, but may be degenerated at one or both ends; lack circular muscles, cilia, excretory system; sexes separate with genital ducts opening into common duct with gut (cloaca).

Genera include: *Gordius*, *Nectonema*, *Paragordius*.

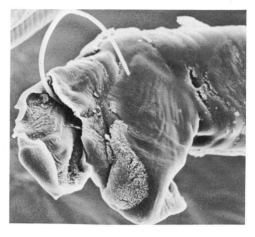

THE roundworm of humans affects some one thousand million people in the world; elephantiasis is a disfiguring disease of the tropics inflicted upon over 250 million individuals. These diseases are just two of the many caused by **roundworms** (Nematoda).

The nematodes are among the most successful groups of animal. They have exploited all forms of aquatic environment and many live in damp soil. Some are even hardy enough to survive in hostile environments like hot springs, deserts and cider vinegar. Furthermore, by parasitizing both animals and plants they have widely extended their habitat ranges. Thus they are significant food and crop pests and the agents of disease in plants, animals and man.

There are many types of nematode, but these animals are all quite similar. They are typically worm-shaped. Their bodies are not divided into segments or regions and generally taper gradually toward both the front and back. The mouth lies at the front and the anus almost at the tail. In cross section, they are perfectly circular and their outer skin is protected by a highly impermeable and resistant cuticle—the secret of the success of this group. The cuticle is complex in structure and is made up of several layers of different chemical composition, each with a different structural layout. The precise form of the cuticle varies from species to species, but in the common gut parasite of man and domestic animals, *Ascaris*, it comprises an outermost keratinized layer, a thick middle layer and a basal layer of three strata of collagen fibers which cross each other obliquely. These flex with respect to each other and allow the animal to make its typical wave-like movements, brought about by longitudinal muscle fibers. The high pressure of fluid maintained in the body cavity (pseudocoel) counteracts the muscles and keeps the worms' cross section round at all times.

Both free-living and parasitic forms may have elaborate mouth structures associated with their food-procuring activities and ways of life. The gut has to function against the fluid pressure of the pseudocoel and in order to do so may depend on a system of pumping bulbs and valves.

In primitive forms excretion is probably carried out by gland cells on the lower surface located near the junction of the pharynx and the intestine. In the more advanced types, there is an H-shaped system of tubular canals with the connection to the excretory pore situated in the middle of the transverse canal.

The nervous system provides a very simple brain encircling the foregut and supplying nerves forward to the lips around the mouth. Other nerves are supplied down the length of the body. The body bears external sensory bristles and papillae.

Free-living nematodes are mostly carnivores feeding on small invertebrates

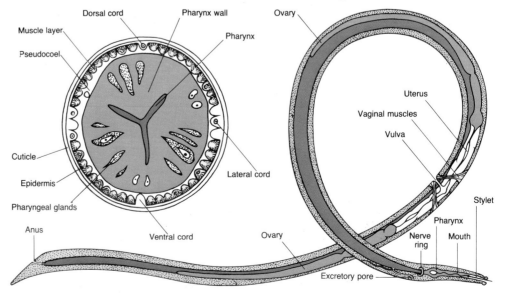

◄ **Pernicious parasite** of man, the hookworm. Shown here is the mating aparatus at the rear end of a male hookworm of the genus *Ancylostoma*. The thread-like protrusion is one of a pair of spicules used to guide sperm into the vulva of the female.

► **Squirming worm**—a free-living nematode worm found in fresh water. The extremely flexible body is bounded by the strong transparent cuticle that is key to the group's success.

▼ **Elephant-like skin** of advanced infections gives elephantiasis its name. The disease, of warm, humid regions of the world, is caused by roundworms (*Wucheria* species). The adult male (1) and female (2) worm parasitize the lymph ducts of their victim, to which they attach themselves, the female by papillae (3) and the male by a spine (4). The adults produce larvae called microfilariae (6) which are sucked up from the blood of an infected person by mosquitoes (5) when they feed. Inside the insect the microfiliariae develop into infective larvae (7) which are injected into the blood of another person when the mosquito feeds. There they move to the lymphatic system, completing the cycle. Adults form the tangled masses of worms that cause the build-up of lymphatic fluids resulting in the characteristic swellings (8).

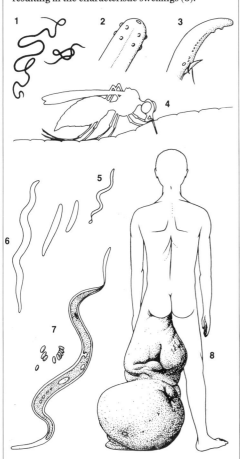

◄ **Anatomy of a roundworm.** (1) Whole animal. (2) Cross section in the region of the pharynx.

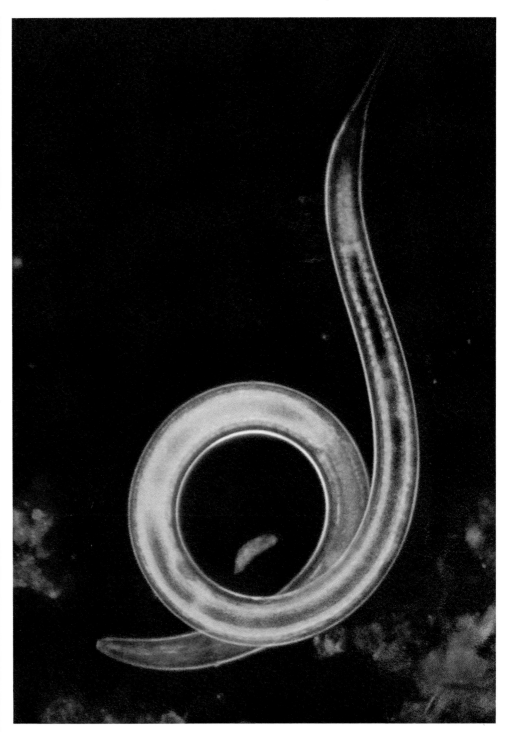

including other nematodes. Some aquatic forms feed on algae and some are specialized for piercing the cells of plant tissue and sucking out the contents. Others are specialized to consume decomposing organic matter such as dung and detritus and/or the bacteria that are feeding on these substances. Nematodes parasitic in man and animals are responsible for many diseases and are therefore of considerable medical and veterinary significance. In humans among other diseases they cause river blindness (*Onchocera* species), roundworm (*Ascaris lumbricoides*) and elephantiasis (*Wucheria* species). Very spectacular is the eye worm *Loa* which can sit on the cornea. Nematode diseases are more common in the tropics.

Parasitic nematodes feed on a variety of body tissues and fluids which they obtain directly from their hosts.

Roundworm of man is one of the largest and most widely distributed of human parasites. Infections are common in many parts of the world and the incidence of parasitism in populations may exceed 70 percent; children are particularly vulnerable to infection. The adult parasites are stout, cream-colored worms that may reach a length of

1ft (30cm) or more. They live in the small intestine, lying freely in the cavity of the gut and maintaining their position against peristalsis (the rhythmic contraction of the intestine) by active muscular movements.

The adult female has a tremendous reproductive capacity and it has been estimated that one individual can lay 200,000 eggs per day. These have thick, protective shells, do not develop until they have been passed out of the intestine and for successful development of the infective larvae a warm, humid environment is necessary. The infective larva can survive in moist soil for a considerable period of time (perhaps years), protected by the shell. Man becomes infected by accidentally swallowing such eggs, often with contaminated food or from unclean hands. The eggs hatch in the small intestine and the larvae undergo an involved migration around the body before returning to the intestine to mature. After penetrating the wall of the intestine the larvae enter a blood vessel and are carried in the bloodstream to the liver and thence to the heart and lungs. In the latter they break out of the blood capillaries, move through the lungs to the bronchi, are carried up the trachea, swallowed and thus return to the alimentary canal. During this migration, which may take about a week, the larvae molt twice. The final molt is completed in the intestine and the worms become mature in about two months.

An infected person may harbor one or two adults only and, as the worms feed largely on the food present in the intestine, will not be greatly troubled, unless the worms move from the intestine into other parts of the body. Large numbers of adults, however, give rise to a number of symptoms and may physically block the intestine. As in trichinosis, migration of the larvae round the body is a dangerous phase in the life cycle and, where large numbers of eggs are swallowed, severe and possibly fatal damage to the liver and lungs may result. Chronic infection, particularly in children, may retard mental and physical development.

Elephantiasis is a disfiguring disease of man restricted to warm, humid regions of the world and occurs in coastal Africa and Asia, the Pacific and in South America.

The blood of humans infected with the parasite contains the microfilaria stage of the worm, that is, the fully developed embryos still within their thin, flexible eggshells, that have been released from the mature female worms. During the day the microfilariae accumulate in the blood vessels of the lungs, but at night, when the

mosquitoes are feeding, the microfilariae appear in the surface blood vessels of the skin and can be taken up by the insects as they suck blood. The daily appearance and disappearance of the microfilariae in the peripheral blood is controlled by the activity pattern of the infected person and is reversed when the person is active at night and asleep during the day.

It is an impressive example of the evolution of close interrelationships between parasites and their hosts and ensures maximum opportunity for the parasite to complete its life cycle. Microfilariae that are taken up by a mosquito undergo a period of development in the body muscles of the insect before becoming infective to man. As the mosquito feeds, larvae may once again enter the human host.

The adult worms may reach a length of 4in (10cm), but are very slender. They live in the lymphatic system of the body, often forming tangled masses and their presence may cause recurrent fevers and pains.

In long-standing infections, however, far more severe effects may be seen, brought about by a combination of allergic reactions to the worms and the effects of mechanical

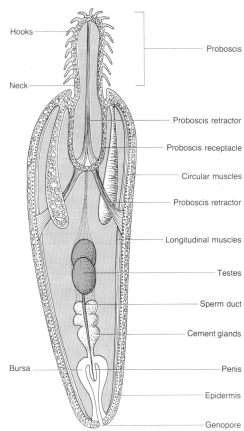

▲ **"Gordian knot"** of massed horsehair worms (*Gordius* species) under a stone in a drying streambed. These worms are often found in drinking troughs used by horses, and for this reason gave rise to the myth that they were horsehair come to life.

◄ **Structure of a spiny-headed worm,** showing the eversible pharynx.

blockage causing accumulation of lymph in the tissues. Certain regions of the body are more commonly affected than others, notably the limbs, breasts, genitals and certain internal organs, which become swollen and enlarged. The skin in these areas becomes thickened and dry and eventually resembles that of an elephant (hence elephantiasis). In severe cases the affected organ may reach an enormous size and thus bring about debilitating or even fatal secondary complications.

Drug treatment for the elimination of the worms is useful in the early stages of infection, but little can be done where chronic disease has produced true elephantiasis. Indeed, the parasites may no longer be present at that time.

Spiny-headed worms are all parasites living in the intestines of various groups of vertebrates. They are particularly successful at parasitizing the bony fish and the birds—they have completely failed to conquer the cartilaginous fishes (skates and rays). Unlike the parasitic flatworms (flukes and tapeworms) and the parasitic roundworms (nematodes), spiny-headed worms are of little medical or economic significance. This is probably because the insects and crustaceans that serve as secondary hosts to the juvenile spiny-headed worms are not eaten by man. In a few species another vertebrate may serve as a secondary host.

Spiny-headed worms completely lack a gut. Food digested by the host's gastric system is absorbed across the body wall and nourishes them. The body wall is composed of a fibrous epidermis which contains channels, sometimes referred to as lacunae, which are not connected to the interior or the exterior of the animal. They probably function to circulate absorbed food materials. The front is equipped with a strange reversible proboscis clad with hooks. The proboscis is extended to attach the worm to the lining of the host's gut. It is retracted by muscles into a special proboscis sac and everted by fluid pressure by a reduction in the proboscis sac volume. Beside the sac are two bodies called lemnisci, filled with small spaces and thought to be food storage areas. There are excretory nephridia in some spiny-headed worms. The nervous system is simple; there is a ventral mass of nerve cells at the front of the body from which arise longitudinal nerve cords.

The sexes are separate and male spiny-headed worms are often larger than females. Internal fertilization takes place with the male using a penis to transfer sperm to the female. The fertilized eggs develop within the female pseudocoel until they reach a larval stage encased in a shell. These larvae are liberated and pass out with the feces of the host. If eaten by the appropriate secondary host, they emerge from the egg cases, penetrate the secondary host's gut wall and come to lie in its blood space where they remain until the secondary host is eaten by the primary host in a prey/predator relationship or by casual accident, as when water containing a small infected crustacean is drunk. At this point the primary host is reinfected.

Horsehair worms are unusual animals closely related to the nematode worms. Adults typically live in the soil around ponds and streams and lay their eggs in the water, attaching them to water plants. The larvae are parasitic and usually attack insects. One of the puzzles about their life cycle is that many of the insects acting as hosts for the horsehair worm larvae are terrestrial, not aquatic, for example, crickets, grasshoppers and cockroaches. It could be that these animals become infected by drinking water containing the larvae. The development inside the host can take several months and the larvae usually leave the host insect when it is near water to lead a free existence in the moist soil. The adults probably do not feed.   AC

Eelworm Traps

In a reversal of the roles played by certain gnat larvae and the fungal fruiting bodies on which they feed above ground, soil-living nematodes face the hazard of predatory fungi. Eelworms, active forms that thread their way through the soil particles, fall prey to over 50 species of fungi whose hyphal threads penetrate the eelworms' bodies.

There are several ways by which the hyphae penetrate the eelworms. Some fungi form sticky cysts which adhere to the eelworm, then germinate and enter its body.

Other fungi form sticky threads and networks, like a spider's web, which trap the eelworms.

Some fungi form lasso-like traps. These consist of three cells forming a ring on a side branch of a hypha. An eelworm may merely push its way into the ring and become wedged or the trap may be "sprung," the three cells suddenly expanding inward in a fraction of a second, to secure the eelworm (see RIGHT).

It is interesting that these fungi do not need to feed on eelworms: they only develop traps if eelworms are present in the soil.

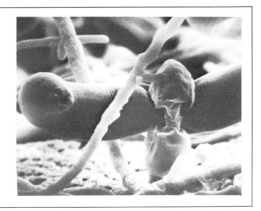

SEGMENTED WORMS

Phylum: Annelida

About 11,500 species in 3 classes.

Distribution: worldwide in land, water and moist habitats.
Fossil record: polychaete tubes (burrows) known from Precambrian (over 600m years ago).
Size: 0.04in–10ft (1mm–3m) long.
Features: body typically elongate, divided into segments and bilaterally symmetrical; body cavity (coelom) present; gut runs full length of body from mouth to anus; nerve cord present in lower part of body cavity; excretion via paired nephridia; appendages, when present, never jointed.

Earthworms
Class: Clitellata

Ragworms and Lugworms
Class: Polychaeta

Leeches
Class: Clitella

▶ **Fan on the sea floor.** Many annelid worms, such as this polychaete tube worm *Serpula vermicularis*, live a sedentary life in a tube, only the fan-like feeding parts protruding to extract food and oxygen from the sea water. In the center of the fan is the round operculum, a type of stopper which is used to close off the tube after the fan is withdrawn.

▼ **Structure of an earthworm,** cut away to show internal structure, particularly the digestive system, reproductive organs and nerve and blood systems.

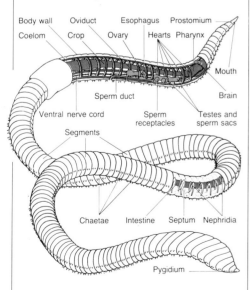

Body wall · Oviduct · Esophagus · Prostomium · Coelom · Crop · Ovary · Hearts · Pharynx · Mouth · Sperm duct · Brain · Ventral nerve cord · Sperm receptacles · Testes and sperm sacs · Segments · Chaetae · Intestine · Septum · Nephridia · Pygidium

▶ **Mating earthworms.** Earthworms (*Lumbricus terrestris*) are hermaphrodite and during mating mutually exchange sperm.

Worms that can be 10ft (3m) long; worms that can suck blood and worms that are vaguely reminiscent of a strip of rag—earthworms, leeches and ragworms. These are representatives of the three groups of worms known collectively as annelids. Annelids are characterized by a long, soft body and cylindrical or somewhat flattened cross-section.

An annelid worm is basically a fluid-filled cylinder, with the body wall comprising two sets of muscles, one circular, the other longitudinal. The fluid-filled cavity (the coelom) is effectively a hydrostatic skeleton (fluid cannot be compressed) upon which the muscles work antagonistically to produce changes in width and length of the animal.

Annelids are divided into a number of segments which in some forms give rise to a ringed appearance of the body—hence the name ringed worms sometimes applied to this group. Although some annelids can be broken up into individual segments and each of these can regenerate from the front or back into a completely new worm, the segments of which the normal adult is composed are not really independent. The gut, vascular system and nervous system run from one end of the segmental chain to the other and coordinate the whole body. The series of segments is bounded by a non-segmental structure at each end: the prostomium and a pygidium. The mouth opens into the gut immediately behind the prostomium, while the gut terminates at the anus on the pygidium. The prostomium and the pygidium are variously adapted: the prostomium, lying at the front, naturally develops organs of special sense such as the eyes and tentacles and hence contains a ganglion or "primitive brain," which connects with the nerve cord that runs from one end of the body to the other. The segments too, although primitively similar, are variously adapted, and the annelids demonstrate the variety of changes that can be rung on this apparently simple and monotonous body plan.

Earthworms and other oligochaete worms have a worldwide distribution. Earthworms may be found in almost any soil, often in very large numbers, and sometimes reaching a great size (*Megascolecides* of Australia reaches 10ft (3m) or more in length). The

Darwin and Earthworms

Charles Darwin, the founder of modern evolutionary theory, published in 1881 a book entitled *The Formation of Vegetable Mould through the Action of Worms*, summarizing 40 years of his observation and experiment. Having discussed the senses, habits, diet and digestion of earthworms, he concluded that they live in burrow systems, feeding on decaying animal and plant material, in addition to quantities of soil. They produce casts, forming a layer of vegetable mold, thereby enriching the surface soil.

He noted that earthworms frequently plug their burrow entrances with either small stones or leaves, the latter also serving as a food source. Darwin saw signs of intelligence in the way in which leaves are grasped. When the leaf tip is more acutely pointed than the base, most are dragged by the tip, but where the reverse is true, the base is more likely to be grasped. Pine needles are almost invariably pulled by the base, where the two elements of the needle are united, forming the "point".

The habit of producing casts at the burrow entrance results in the gradual accumulation of surface mold, and Darwin estimated that annually some 8–18 tons of soil per acre (20–45 tons per hectare) may be brought to the surface in pasture. This soon buries objects left on the soil surface, and may aid the preservation of archaeological sites, but under certain conditions cast formation may also accelerate soil erosion. Burrowing affects the aeration and drainage of the soil, which is beneficial to crop production.

Darwin failed to distinguish between earthworm species (only relatively few British earthworms make permanent burrows or produce surface casts), but he was one of the first to recognize the important role that earthworms play in the terrestrial habitat. He firmly established their beneficial effects on soil, which have subsequently been largely confirmed.

commonest European earthworms are *Lumbricus* and *Allolobophora* species. The aquatic oligochaetes, sometimes known as bloodworms because of their deep red color, are smaller than earthworms and generally simpler in structure. Some are found in the intertidal zone, under stones or among seaweeds. But many of the aquatic genera are found in freshwater habitats, living, for example, in mud at the bottom of lakes (*Tubifex*). Such worms commonly have both anatomical and physiological adaptations which equip them to withstand the relatively deoxygenated conditions commonly found in polluted habitats.

Oligochaetes are remarkably uniform. They entirely lack appendages, except for gills in a few species. Some families are restricted to one type of environment, for example where there are fungi and bacteria associated with the breakdown of organic material; some may eat just microflora.

Oligochaetes are exclusively hermaphrodite, with reproductive organs limited to a few front segments. Male and female segments are separate, with segments containing testes always in front of those with ovaries (the reverse is true in leeches). The clitellum, a glandular region of the epidermis (outer skin), is always present in mature animals. It secretes mucus to bind together copulating earthworms, and produces both the egg cocoon and the nutritive fluid it contains. The exact position of the

The 3 Classes of Segmented Worms

Earthworms

Class: Clitellata, order Oligochaeta
About 3,000 species in 284 genera.

Distribution: worldwide in terrestrial, freshwater, estuarine and marine habitats.

Size: 0.04in–10ft (1mm–3m) long.

Features: no head appendages; few bristle-like chaetae usually in 4 bundles per segment; coelom spacious and compartmented; has blood vascular system, well-developed body-wall musculature and ventral nerve cord with giant fibers in some; bisexual with reproductive organs confined to a few segments; glandular saddle (clitellum) present which secretes cocoon in which eggs develop; no larval stage.

Families: Aeolosomatidae, 4 genera; Alluroididae, 4 genera; Dorydrilidae, 1 genus; Enchytraeidae, 23 genera, including *Enchytraeus*, *Marionina*, *Achaetus*; Eudrilidae, 40 genera; Glossoscolecidae, 34 genera; Haplotaxidae, 1 genus; Lumbricidae, 10 genera, including **earthworms** (*Lumbricus, Allolobophora, Eisenia*); Lumbriculidae, 12 genera; Megascolecidae, 101 genera, including *Megascolex, Pheretima, Dichogaster*; Moniligastridae, 5 genera; Naididae, 20 genera, including *Chaetogaster, Dero, Nais*; Opistocystidae, 1 genus; Phreodrilidae, 1 genus; Tubificidae, 27 genera, including some **bloodworms** (*Tubifex* species), *Peloscolex*.

Ragworms and lugworms

Class: Polychaeta
About 8,000 species in some 80 families.

Distribution: worldwide, essentially marine.

Size: 0.04in–6.6ft (1mm–2m) long.

Features: morphology extremely variable; various feeding and sensory structures may be present at front end; body segments with lateral appendages (parapodia) bearing bristle-like chaetae; coelom spacious; has blood vascular system, nephridia and ventral nerve cord with giant nerve fibers in some; usually sexual reproduction, with or without free-swimming larval phase.

About 80 families including: Aphroditidae (**sea mice**); Arenicolidae, including **lugworm** (*Arenicola marina*); Cirrotulidae; Eunicidae; Glyceridae, including some **bloodworms** (*Glyceris* species); Nephthyidae, including **catworms** (*Nephthys* species); Nereididae, including **ragworm** or **sandworm** (*Nereis virens*); Onuphidae (**beachworms**); Opheliidae; Polynoidae (**scaleworms**); Sabellariidae; Sabellidae (**fanworms**); Serpulidae; Spionidae; Syllidae; Terebellidae (including some **bloodworms**).

Leeches

Class: Clitella, order Hirudinea
About 500 species in 140 genera.

Distribution: worldwide, predominantly freshwater, some terrestrial or marine.

Size: 0.2–4.7in (5mm–12cm) long.

Features: number of segments 33; suckers at front and rear; coelom greatly reduced; blood vascular system tends to be restricted to remaining coelomic spaces (sinuses); hermaphrodite with one of mating pair transferring sperm; glandular saddle (clitellum) produces cocoon; no larval stage.

Families: Acanthobdellidae, genus *Acanthobdella*; Americobdellidae, genus *Americobdella*; Erpobdellidae, including *Erpobdella*; Glossiphoniidae, including *Glossiphonia, Theroemyzon, Placobdella*; Haemadipsidae, including *Haemadipsa*; Hirudidae, including **Medicinal leech** (*Hirudo medicinalis*), *Haemopis*; Piscicolidae, including *Pisciola, Branchellion, Ozobranchus*; Semiscolecidae, including *Semiscolex*; Trematobdellidae, including *Trematobdella*; Xerobdellidae, including *Xerobdella*.

clitellum and the details of the reproductive organs and their exterior openings is of fundamental importance in the classification and identification of oligochaetes.

During copulation two earthworms come together, head to tail, and each exchanges sperm with the other. The sperm are stored by each recipient in pouches called spermathecae until after the worms separate. A slimy tube formed by the clitellum later slips off each worm, collecting eggs and the deposited sperm as it goes, and is left in the soil as a sealed cocoon. The eggs are well supplied with yolk, and are also sustained by the albumenous fluid surrounding them. In earthworms there is a tendency for the eggs to be provided with less yolk and rely more on the albumen. Development of the worms is direct, that is, there are no larval stages.

Parthenogenesis (development of unfertilized eggs) and self-fertilization are known in some species, and the aquatic members of the Aeolosomatidae and Naididae almost exclusively reproduce asexually.

Some earthworms are potentially long-lived, but the majority of oligochaetes probably have one- to two-year life cycles. Cocoon production occupies much of the year, interrupted by unfavorable environmental conditions, such as dryness, or by a seemingly fixed dormancy in some earthworms, that does not appear to be related to outside conditions.

An earthworm moves by waves of muscular contraction and relaxation which pass along the length of the body, so that a particular region is alternately thin and extended or shortened and thickened. A good grip on the walls of the burrow is aided by the spiny outgrowths of the body wall called chaetae which project from the body wall. Chaetae are found in both polychaetes and oligochaetes; but, as their names imply, polychaetes have many chaetae in each segment, whereas oligochaetes have relatively few. Furthermore, the number of chaetae in each segment of an oligochaete is not only small but constant. The earthworm *Lumbricus*, for example, has eight chaetae in each segment. In the aquatic oligochaetes the chaetae may be longer and more slender than those of an earthworm.

The earthworm feeds either on leaves pulled into the burrow with the aid of its suctorial pharynx, or by digesting the organic matter present among the particles of soil which it swallows when burrowing in earth otherwise too firm to penetrate. A muscular gizzard near the front end of the gut serves to break up compacted soil particles into

Bloodsucker—the Medicinal Leech

The most familiar of leeches is the Medicinal leech, which is a native of Europe and parts of Asia. Its use in blood letting in the 18th and 19th centuries led to its introduction to North America. However, it is now believed to be extinct there, and it has also recently become extinct in Ireland. The medicinal value of the practice of "blood letting" for almost all imaginable ills, "vapors and humors" may be questioned, but there is no doubt that the leech's ability to gorge fast for long periods is a perfect adaptation to its life-style.

The Medicinal leech eats blood, usually mammalian but also that of amphibians and even fish. Once attached to a potential blood source by means of its suckers, its three jaws are brought into contact with the skin. Each is shaped like a semicircular saw, bearing numerous small teeth. Sawing action of the teeth results in a Y-shaped incision through which blood is drawn by the sucking action of the muscular pharynx. Glandular secretions are released through each tooth which dilate the host's blood vessels, prevent coagulation of the blood, and act as an anesthetic.

A feeding Medicinal leech may take up to five times its body weight in blood, which passes into its spacious crop. Water and inorganic substances are rapidly extracted and excreted through the excretory cells

(nephridia), but digestion of the organic portion may take some 30 weeks. It was once thought that the Medicinal leech entirely lacked digestive enzymes, relying on bacteria in its own gut for digestion. However, although the involvement of bacteria is still recognized the leech does produce its own enzymes. The time taken to digest this highly nutritious food means that leeches need to feed only infrequently—it has been suggested that not much more than one full meal per year would be sufficient to permit growth.

Although little is known of its life history, the Medicinal leech is thought to take at least three years to mature. Cocoons, containing 5–15 eggs, are laid in damp places 1–9 months after copulation; they hatch 4–10 weeks later.

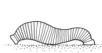

smaller ones, with the result that digestion and absorption of the organic material within the intestine is more efficient. Undigested matter is extruded from the anus on to the surface of the soil as the familiar worm casts. These casts give some indication of the valuable effects earthworms have upon the soil (see box, p70).

The best known of all **leeches** is the Medicinal leech which for centuries has been used in the "medicinal practice" of blood letting (see box). Leeches are easily recognized. Their soft bodies are ringed, usually without external projections and with a prominent often circular sucker at the rear. At the front there is another sucker around the mouth, and although this may be quite prominent as in fish leeches, it frequently is not. The skin is covered with a thin cuticle onto which mucus is liberally secreted. In front there is usually a series of paired eyes.

Leeches lack chaetae (except in the Acanhobdellidae) and have a fixed number (33) of segments. Like oligochaetes, they possess a clitellum, involved in egg cocoon production.

Leeches are divisible into two types: those with an eversible proboscis (Glossiphoniidae and Pisciolidae) and those with a muscular sucking pharynx, which may be unarmed (Erpobdellidae, Trematobdellidae, Americobdellidae, Xerobdellidae) or armed with jaws (Hirudidae, Haemadipsidae).

Not all species feed in the manner of the Medicinal leech. Some species feed on other invertebrates and may either suck their body fluids (eg Glossiphoniidae) or swallow them whole (eg Erpobdellidae). The Hirundidae, including the Medicinal leech, and the terrestrial Haemadipsidae feed on vertebrate blood, often that of mammals. The Pisciolidae are mostly marine, living on fish body fluids.

As well as circular and longitudinal muscles in the body wall, diagonal muscle bundles form two systems spiralling in opposite directions between these layers and opposing both. Muscles running from top to underside are well developed and are used to flatten the body, especially during swimming and movements associated with respiration.

The coelomic space is largely invaded by tissue, leaving only a system of smaller spaces (sinuses). The internal funnels of the excretory nephridia open into this system and may be closely associated with the blood system, playing a part in blood fluid production. The blood vascular system is intimately connected with the coelomic sinuses, and may be completely replaced by them.

At copulation, one animal normally acts as sperm donor, and in the jawed leeches, a muscular pouch (atrium) opening to the outside acts as a penis for sperm transfer to the female gonadal pore. In other leeches, sperm packets formed in the atrium attach to the body wall of the recipient, and the spermatozoa make their way through the body tissues by a poorly understood mechanism. Fertilization is internal or occurs as gametes are released into the cocoon, which is secreted by the clitellum and slipped over the leech's head. A small number of zygotes are put in each cocoon, which is full of nutrient fluid secreted by the clitellum. In the Glossiphoniidae, cocoons are protected by the parent, and the juveniles may spend some time attached to the parent. Most leeches only go through one or two breeding periods, with one- to two-year life cycles.

Leeches are generally accepted as having been derived from oligochaetes, specializing as predators or external parasites. The fish parasite *Acanthobdella* shows many features intermediate between the two groups.

In contrast to the other annelids, polychaetes, such as **ragworms**, have extremely diverse forms and biology, although a common pattern is usually recognizable for each family. Polychaetes are almost all marine and they are often a dominant group, from the intertidal zone to the depths of the oceans. Planktonic, commensal and parasitic forms are also found.

Basically a polychaete is composed of a series of body segments, each separated from its neighbor by partitions (septa). Externally, each segment bears a pair of bilobed muscular extensions of the body wall known as parapodia containing internal supports (acicula), two bundles of chaetae, and a pair of sensory tentacles (cirri). The parapodia are best developed in active crawling forms, for example, ragworms.

Many polychaete families have an eversible pharynx which is most conspicuous in those species that are carnivorous, feed on large pieces of plant material or suck body fluids of other organisms. The pharynx in such families may be armed with jaws. Families with feeding tentacles or crowns use

▲ **Leech locomotion.** The suckers are attached alternately to the substrate, lifted and moved forward, thus removing the need for chaetae and a fluid-filled coelom, both of which are characteristic features of other annelid worms.

◄ **Waiting to pounce** upon passing prey, a jawed leech in the leaf litter of the north Australian rain forest. One good meal of blood each year may be enough for nutritional needs.

◄ **Taking a blood meal** OPPOSITE BELOW, a Medicinal leech hangs onto a human arm. A Medicinal leech may take up to five times its body weight in blood at each meal.

▼ **Structure of a leech.** Externally the body is divided into numerous "segments," although anatomically only 33 (34 according to some) are present, including those fused to form the two suckers. Leeches are protandrous hermaphrodites, with one pair of ovaries and 4–10 pairs of testes, each gonad lying in a sac in the body cavity (coelom). The testes are linked via a system of ducts to a muscular chamber, the atrium, opening to the exterior. The ovaries are positioned in front of the first pair of testes.

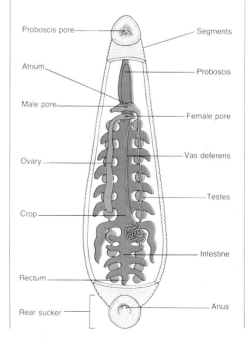

Proboscis pore — Segments
Atrium — Proboscis
Male pore — Female pore
Ovary — Vas deferens
Crop — Testes
Rectum — Intestine
Rear sucker — Anus

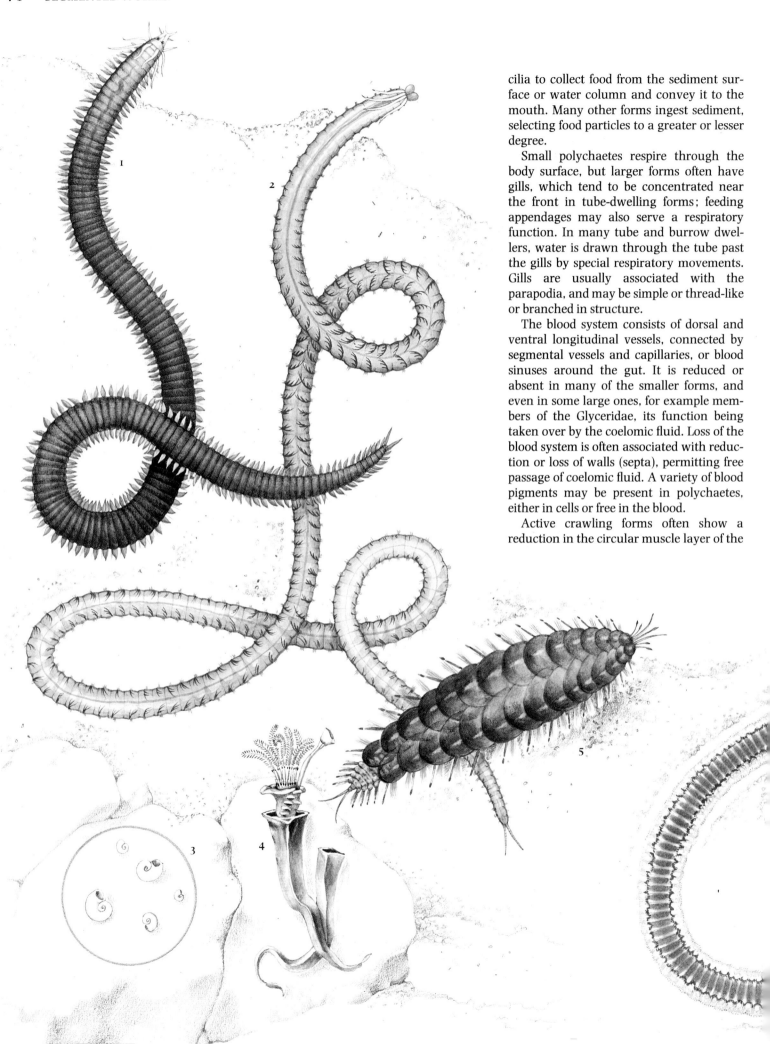

cilia to collect food from the sediment surface or water column and convey it to the mouth. Many other forms ingest sediment, selecting food particles to a greater or lesser degree.

Small polychaetes respire through the body surface, but larger forms often have gills, which tend to be concentrated near the front in tube-dwelling forms; feeding appendages may also serve a respiratory function. In many tube and burrow dwellers, water is drawn through the tube past the gills by special respiratory movements. Gills are usually associated with the parapodia, and may be simple or thread-like or branched in structure.

The blood system consists of dorsal and ventral longitudinal vessels, connected by segmental vessels and capillaries, or blood sinuses around the gut. It is reduced or absent in many of the smaller forms, and even in some large ones, for example members of the Glyceridae, its function being taken over by the coelomic fluid. Loss of the blood system is often associated with reduction or loss of walls (septa), permitting free passage of coelomic fluid. A variety of blood pigments may be present in polychaetes, either in cells or free in the blood.

Active crawling forms often show a reduction in the circular muscle layer of the body wall with corresponding development of the parapodia and their musculature and reduction or loss of the septa. This is correlated with a switch from peristaltic locomotion (as in earthworms) to the use of parapodia as the main propulsive organs.

Connecting the coelomic fluid to the exterior are a pair of excretory nephridia and a pair of genital ducts, the coelomoducts, in each segment. The nephridia pass through a septum, the external pore occurring in the segment behind that in which the nephridium originates. In most polychaetes the coelomoducts and nephridia are fused in fertile segments to produce urinogenital ducts, although often only at sexual maturity. Fertilization is generally external, and may lead to a free-swimming larva which may or may not feed, or development may be entirely on the sea floor (benthic). Adults protect the brood in a number of species, and this is not restricted to those with benthic development. Polychaetes may be hermaphrodite, eggs may develop into larva within the body (viviparous) or internal fertilization may take place. Asexual reproduction, by fragmentation or fission followed by regeneration, occurs in some 30 species.

Ragworms are perhaps the most familiar

◀ ▼ **Representative species of polychaet worms,** a major element in the benthic fauna of the world's seas. (**1**) *Eulalia viridis*, N Atlantic (6in, 15cm). (**2**) *Marphysa sanguinea*, N Atlantic (24in, 60cm). (**3**) *Spirorbis borealis*, a fanworm, N Atlantic (0.1in, 2.5mm). (**4**) *Pomatoceros triqueter*, a fanworm, N Atlantic (1in, 2.5cm). (**5**) *Harmothoe inibricata*, a scaleworm, N Atlantic (2in, 5cm). (**6**) *Hermodice carunculata*, Mediterranean (12in, 30cm). (**7**) *Perineretis nuntia*, a ragworm, Indo-Pacific (1in, 2.5cm). (**8**) Tentacles of *Reteterebella queenslandica*, Indo-Pacific (1.4in, 3.5cm). (**9**) *Sabellastarte intica*, Indo-Pacific (1in, 2.5cm). (**10**) *Spirobranchus gigantens*, Indo-Pacific (0.6in, 1.5cm).

polychaete worms, mainly through their use by sea anglers. Some species are carnivorous, feeding on dead or dying animals; some are omnivorous; others feed only on weed. Polychaete worms predominate among bait species. In Europe ragworm and lugworm are most commonly used for bait, and the catworms occasionally collected. In North America the sandworm and the bloodworm are extensively used, as are the long beachworms (species of the family Onuphidae) in Australia. All these species live in intertidal soft sediment, although ragworms are often found in muddy gravel.

They live for protection in the mud or muddy sand from which they emerge or partially emerge to feed on the plant and animal debris on the surface. The burrows may be located by the holes on the surface of the mud. The burrows themselves are U-shaped or at least have two openings, for the worms must irrigate them in order to respire. They do this by undulating their bodies, and this serves not only to renew the water within the burrow, but enables them to detect in the incoming water signs of food in the vicinity. Vast populations may occur in suitable habitats. Some species can tolerate the stringent conditions in estuaries and some are found in very low salinities.

Scaleworms also occur on the seashore and below the low water mark. Their backs are covered or partly covered by a series of more or less disk-shaped scales. The scales are commonly dark in color and overlap slightly. They are protective, not only in providing concealment but also in their ability to luminesce. The luminescent organs are under nervous control, so that the animal can "flash" when alarmed. Not all scaleworms do this, however.

Another polychaete worm is the sea mouse which is found below low tide level on sandy bottoms. It is often brought up by fishermen in the dredge. The body is covered with a fine "felt" of silky chaetae, and it is

▲ **Marine mouse.** The fine felt of silky bristles or chaetae gives this polychaete worm a furry appearance from which the common name of sea mouse is derived. Shown here is *Aphrodite aculeata*. Sea mice spend most of their lives in sand or mud in shallow waters.

▶ **Lawn of fanworms** spreading their feeding crowns from the seabed. These representatives of the family Sabellidae are found off Lizard Island, Queensland, Australia at a depth of 100ft (30m).

▼ **Structure of a polychaete worm.** (1) Whole body. (2) Cross section. The basic structure is a series of body segments, each separated from its neighbor by partitions, the septa. Externally, a segment bears a pair of bilobed parapodia containing internal supports, the acicula, two bundles of chaetae, and a pair of sensory cirri. In front of the mouth, which is on a segment that often lacks chaetae, is the prostomium containing the brain.

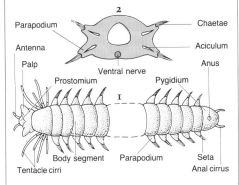

▶ **Varied heads of polychaete worms.** In families where it is well developed, the prostomium may have no appendages or (1) bear a small number of sensory antennae, with or without a pair of sensory/feeding palps; eyes are often present, for example in *Nothria elegans*. In species without such appendages, feeding tentacles may be present behind the prostomium, as in the families Spionidae and Cirratulidae. In other families the prostomium is reduced due to the presence of feeding appendages in the form of extensible tentacles, as in (2) members of the family Terebellidae, or a stiff crown or fan, as in the families Sabellidae (3) and Serpulidae, conditions associated with life in a tube. Specialized chaetae may form part of a lid-like opercular structure in other tube-dwelling families, such as (4) the Sabellariidae.

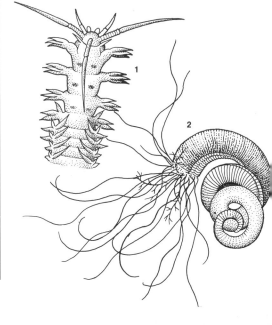

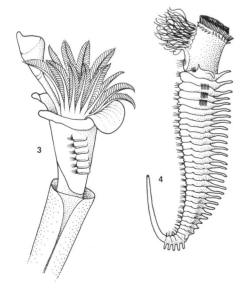

3

4

to this apparently furry appearance that its common name is due.

Sea mice live for the most part just beneath the surface of the sand or mud. They are rather lethargic although they can scuttle quite rapidly for a short distance when disturbed. If the matted "felt" over the back is removed a series of large disk-shaped scales will be seen overlapping the back: the sea mouse is nothing more than a large scaleworm in which the characteristic scales are concealed by the hair-like chaetae.

Fanworms are the most elegant of all the marine polychaete worms. Pinnate or branched filaments radiate from the head to form an almost complete crown of orange, purple, green or a combination of these colors. The crown is developed from the prostomium. It forms a feeding organ and, incidentally, a gill. The remainder of the body is more or less cylindrical.

All fanworms secrete close-fitting tubes which provide protection. Although their often gaily-colored crowns must tempt predatory fish, they can all contract with startling rapidity. These startle responses are made possible by relatively enormous giant nerve fibers which run from one end of the body to the other within the main nerve cord. In *Myxicola* this giant fiber occupies almost the whole of the nerve cord. Almost 0.04in across, it is one of the largest nerve fibers known. Giant nerve fibers are associated with particularly well-developed longitudinal muscles which enable the worm to retract promptly when danger threatens. Other movements are relatively slow.

PRG

ECHIURANS, PRIAPULANS, SIPUNCULANS

Phyla: Echiura, Priapula, Sipuncula

Echiurans
Phylum: Echiura
About 130 species in 34 genera.
Distribution: worldwide, largely marine but with a few brackish water species; intertidal down to 33,000ft (10,000m).
Size: usually between 1.2 and 6in (3–15cm) long, occasionally up to 30in (75cm).

Features: unsegmented worms with coelom; trunk with muscular proboscis at front; mouth at base of proboscis; gut coiled; anus terminal; blood vascular system usually present; unsegmented ventral nerve cord; usually a pair of bristles (setae) just behind mouth; one to many excretory nephridia and a pair of anal excretory organs; sexes separate with extreme sexual dimorphism in some; produce free-swimming trochophore larva in some others.

Families: Bonelliidae, including genus *Bonellia*; Echiuridae, including genera *Echiurus*, *Thalessema*; Ikedaidae, including genus *Ikeda*; Urechidae, including genus *Urechis*.

Priapulans
Phylum: Priapula
Nine species in 6 genera.
Distribution: exclusively marine, intertidal and subtidal.
Size: 0.08–16in (0.2–20cm).

Features: unsegmented worms with coelom; body divided into 2 or 3 regions; introvert at front with terminal mouth and eversible pharynx; trunk may have tail attached; possess chitinous cuticle which is periodically molted; cilia lacking; sexes separate; produce free-living lorica larva.

Families: Priapulidae (genera *Priapulus*, *Priapulopsis*, *Acanthopriapulus*, *Halicryptus*); Tubiluchidae (genus *Tubiluchus*); Chaetostephanidae (genus *Chaetostephanus*).

Sipunculans
Phylum: Sipuncula
About 320 species in 17 genera.
Distribution: worldwide, exclusively marine; intertidal down to 21,000ft (6,500m).
Size: trunk 0.1–20in (0.3–50cm); introvert half to 10 times as long as trunk.

Features: unsegmented worms with coelom; body divided into introvert and trunk; mouth at tip of retractable introvert; gut U-shaped with anus usually at front of trunk; possess 1–2 excretory nephridia; ventral nerve cord unsegmented; trunk wall of outer circular and inner longitudinal muscles; asexual reproduction rare; usually unisexual; larvae free-swimming.

Families: Sipunculidae, including genus *Sipunculus*; Golfingiidae, including genera *Golfingia*, *Phascolion*, *Themiste*; Phascolosomatidae (genera *Phascolosoma*, *Fisherana*); Aspidosiphonidae, including genus *Aspidosiphon*.

ECHIURANS, priapulans and sipunculans are three groups of marine animals whose existence is only really known to scientists—hence they lack common names! All are worm-like, but their bodies are not segmented as in segmented (annelid) worms. The three groups are not particularly closely related to each other.

Echiurans are delicate soft-bodied animals living in soft sediments or under stones in semi-permanent tubes, or inhabiting crevices in rock or coral. They are exclusively marine, although a few forms penetrate into brackish waters, and many members of the family Bonelliidae are found at very great depth.

Food, usually in the form of surface detritus, is caught up in mucus and transported by cilia down the muscular proboscis to the mouth. In *Urechis* species a net of mucus is used to filter bacteria out of water pumped through the burrow, and the net is then consumed along with the food.

In *Urechis* the rear end of the gut is modified for respiration, rhythmic contractions of the hind gut drawing in and expelling water. This may be linked to the absence of a blood circulatory system in this genus, the coelomic fluid containing blood cells having taken over this function. In all other echiurans the blood circulatory system is separate from the coelom.

The musculature of the trunk wall includes an outer circular layer and an inner longitudinal layer. An additional oblique layer may be present, its position being an important feature in defining families.

Reproduction is always sexual, the sexes being separate. Mature gametes are collected into the excretory nephridia just before spawning, in some cases by complex coiled collecting organs. Where fertilization occurs in the water (families Echiuridae, Urechidae) a free-swimming trochophore larva results, gradually changing to the adult morphology. In the Bonelliidae, where the male lives in or on the female, fertilization is presumed to be internal and larval development is essentially on the sea floor. In *Bonellia*, if the larva makes contact with an adult female, it tends to become male, and if not, female.

Priapulans comprise a small phylum of uncertain affinities, composed of exclusively marine species inhabiting soft sediments. The more familiar forms are the relatively large family Priapulidae, but recently the small sand-living genus *Tubiluchus* and the small tube-dwelling *Chaetostephanus* have been described.

The trunk of priapulans bears warts and spines, and although it may have rings, no true segmentation is present. A tail consisting of one or two projections may be present, although there is no tail in *Halicryptus* or *Chaetostephanus*. In *Priapulus* the tail is in the form of a series of vesicles which constantly change shape and volume and may be involved in respiration. In *Acanthopriapulus* the tail is muscular, bearing numerous hooks, and may serve as an anchor during burrowing. In *Tubiluchus* the tail is a long tubular structure.

The pharynx is eversible and muscular, armed with numerous teeth, or, in *Chaetostephanus*, bristles (setae). In the Priapulidae, it is used to capture living prey, but in *Tubiluchus* a scraping function is more likely. The intestine is in the trunk, the anus opening at its rear, with no part of the gut in the tail.

The sexes are separate, with differences in size between the sexes in *Tubiluchus*, probably associated with copulation.

The free-living larval stage may be extremely long, the larva feeding in surface sediment layers for two years or more before changing to the adult condition.

The Priapula were formerly considered to be related to the Kinorhyncha (p50) because of the body armature, lack of cilia and the molting of the cuticle. Now, however, they are regarded as a coelomate group.

Sipunculans occupy a variety of marine habitats. Many live in temporary burrows in soft sediment, while others can bore into chalky (calcareous) rocks, as in the genera *Phascolosoma* and *Aspidosiphon*, or corals, as in *Cloeosiphon* and *Lithacrosiphon*. Some, such as *Phascolion*, inhabit empty gastropod shells.

All sipunculans have an extensible introvert up to 10 times the length of the trunk, with the mouth at its tip, which can be withdrawn completely into the trunk. It allows feeding at the substrate surface while the body remains protected, and can be readily regenerated if lost. The introvert is extended by fluid pressure brought about by contraction of the musculature of the trunk wall. In some cases, for example *Sipunculus* species, the circular and longitudinal muscle layers may be arranged in separate bundles. Up to four special muscles retract the introvert. Most sipunculans consume sediment, although some are thought to filter material from the water using cilia on their tentacles.

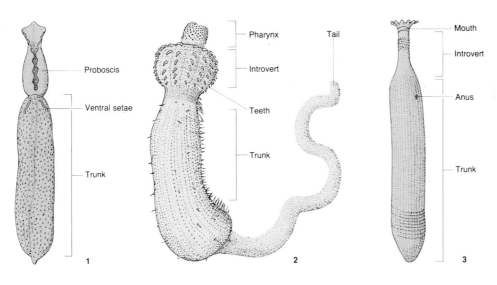

Pharynx

Tail

Mouth

Proboscis

Introvert

Introvert

Ventral setae

Teeth

Anus

Trunk

Trunk

Trunk

1

2

3

▲ **Body forms.** (1) In echiurans the proboscis is sited in front of the mouth and is not entirely retractable into the trunk. When fully extended, the muscular proboscis may be several times the length of the trunk. It may be forked and is well supplied with cilia and mucus glands.

(2) The priapulan body is divided into 2–3 regions—an introvert at the front, a trunk and a tail (which may be absent). The introvert bears the mouth at its tip and is a large barrel-shaped structure armed with rows of teeth or, in *Chaetostephanus*, with complex setae, and is a characteristic feature of the group. It can be withdrawn into the trunk by a series of muscles and is used in burrowing.

All sipunculans have (3) an extensible introvert up to 10 times the length of the trunk, with the mouth at its tip, which can be withdrawn completely into the trunk. It allows the sipunculan to feed at the surface of the seabed while the body remains below it protected. The introvert of a sipunculan can readily be regenerated if it is lost.

▼ ▶ **Representative species of minor worms.** (1) *Thalassema neptuni*, an echiuran (2.8in, 7cm). (2) *Priapulus caudatus*, a priapulan; N Atlantic (3in, 8cm). (3) *Chaetopterus variopedatus*; N Atlantic and Mediterranean (9in, 23cm). (4) *Phascolion strombi*, a sipunculan which lines empty shells (eg that of *Turitella*) with mud and forms a burrow inside; N Atlantic.

Excretion occurs through the one or two nephridia, which connect the large fluid-filled coelomic space to the exterior, opening on the front of the trunk. They also serve as storage organs for gametes immediately before spawning. Sipunculans have no blood vascular system, the coelomic fluid performing the circulatory function.

Externally, sipunculans may have hooks or spines on the introvert and small protrusions (papillae) and glandular openings on the introvert and trunk. In many boring forms a horny or calcareous shield is present at the front end of the trunk, protecting the animal as it lies retracted in its burrow.

Asexual reproduction is known in only two species, sexual reproduction with separate sexes being the rule. Some populations of *Golfingia minuta* are hermaphrodite. In all sipunculans studied, much of the germ cell development occurs free in the coelomic fluid. At spawning, eggs and sperm are released via the nephridia, and fertilization occurs in the sea. The mode of larval development, however, varies between species and may include a free-swimming trochophore-type larva, or may be direct, with juvenile worms emerging from egg masses. In several species, the floating stage occupies a period of months, and long distance transportation of such larvae across the Atlantic has been suggested.   PRG

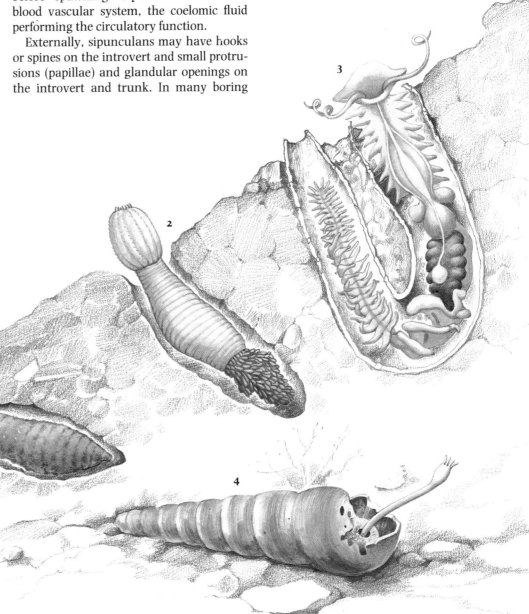

3

2

1

4

BEARD WORMS AND ARROW WORMS

Phyla: Pogonophora and Chaetognatha

Beard worms

Phylum: Pogonophora

About 150 living species in 2 orders and 7 families.

Distribution: all oceans, usually at considerable depths.

Fossil record: none.

Size: length 2in to 5ft (5cm–1.5m), usually 3.2–6in (8–15cm); diameter up to 0.1in (3mm).

Features: bilaterally symmetrical, solitary, tube-dwelling worms; body divided into 3 zones—head (cephalic) lobe bearing tentacles, trunk, and posterior opisthosoma; gill slits, digestive tract and anus all lacking; protostome development.

Order: Athecanephria
Families: Oligobrachiidae (including genus *Oligobrachia*) and Siboglinidae (including genus *Siboglinum*).

Order: Thecanephria
Families: Lamellibrachiidae (including genus *Lamellibrachia*); Lamellisabellidae (including genus *Lamellisabella*); Polybrachiidae; Sclerolinidae (including genus *Sclerolinum*); Spirobrachiidae (including genus *Spirobrachia*).

Arrow worms

Phylum: Chaetognatha

About 70 living species in 7 genera.

Distribution: all oceans, planktonic apart from 1 benthic genus.

Fossil record: one dubious record.

Size: between 0.16in (0.4cm) and 4in (10cm), most less than 0.8in (2cm) long.

Features: small bilaterally symmetrical free-living animals with deuterostome development; lack circulatory and excretory systems; body torpedo-shaped with paired lateral fins and tail fin; mouth at front and armed with strong grasping spines.

Seven genera: *Bathyspadella*, *Eukrohnia*, *Heterokrohnia*, *Krohnitta*, *Pterosagitta*, *Sagitta*, *Spadella*.

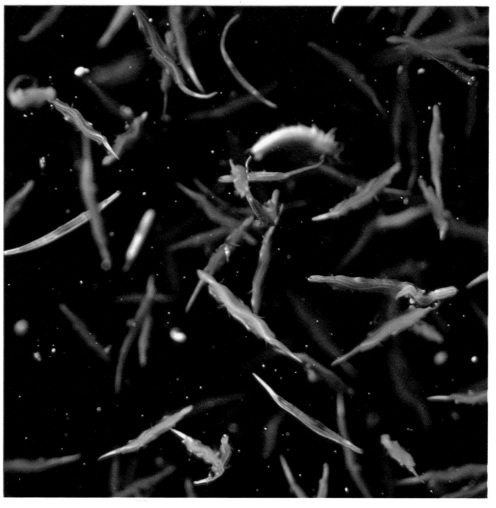

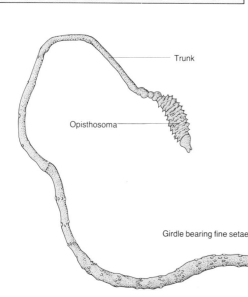

Trunk

Opisthosoma

Girdle bearing fine setae

Papillae

Trunk

BEARD worms are the most unlikely of animals. They are a zoological curiosity of comparatively recent discovery. Their anatomy was not fully appreciated until 1963 when the first whole specimens were obtained. The name Pogonophora comes from the Greek word *Pogon* meaning a beard—this being a reference to the shaggy group of tentacles carried on the front of the body.

These long, thin animals—often 500 times as long as broad—live in tubes which are generally completely buried in the mud or ooze of the ocean floor. There are two exceptions to this: *Sclerolinum brattstromi* lives inside rotten organic matter like paper, wood and leather in Norwegian fjords; and *Lamellibrachia barhami* builds tubes which project from the sediment.

Beard worms have a front head (cephalic) lobe bearing between one and 100 or more tentacles. Immediately behind the head is the forepart, or bridle, which appears to be important in tube building. Behind this lies the trunk, which comprises the greatest part of the body. It is covered with minute tentacles (papillae) which toward the front are arranged in pairs in quite a regular fashion. They become more irregular further back along the trunk and some may be enlarged. These papillae are thought to enable the worm to move inside its tube, and they are associated with plaques, hardened plates which probably assist in this respect. Further along the trunk there is a girdle where the skin is ridged and rows of bristles, similar to those of annelid worms, occur on the ridges. These probably help the worm maintain its position inside the tube.

The distinct rear region of the body, the opisthosoma, is easily broken from the rest of the body, and because of this its presence was not appreciated for many years. It comprises between five and 23 segments, each of which carries bristles larger than the ones on the girdle. The opisthosoma probably

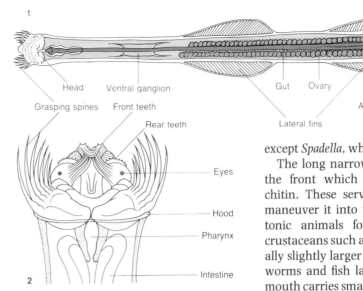

1

Head   Ventral ganglion   Gut   Ovary   Anus   Seminal vesicles

Grasping spines   Front teeth   Lateral fins   Testes   Tail fin

Rear teeth

Eyes

Hood

Pharynx

Intestine

2

▶ **Body plan of an arrow worm.** (1) Whole body, showing the main internal organs. (2) Detail of head, showing spines and teeth used in feeding.

◀ **Arrow worms** in marine plankton. The existence of this group of animals was not discovered until 1829, when nets capable of sieving out plankton from the sea were developed.

▼ **Beard worm body plan.** A typical pogonophoran, showing the whole worm, most of which is normally buried in its narrow tube in the seabed.

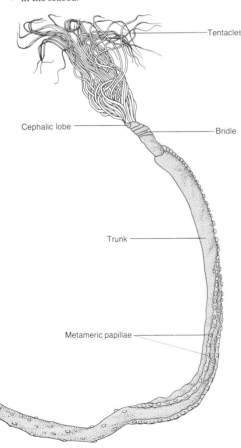

Tentacles

Cephalic lobe

Bridle

Trunk

Metameric papillae

protrudes from the bottom of the tube in life and acts as a burrowing organ.

The tube itself is made from chitin, and according to the species the lengths range from a few inches to about 5ft (1.5m). Beard worm tubes are however, very narrow, never exceeding 0.1in (3mm) in diameter. They are quite stiff and may appear banded.

One of the most peculiar features of the beard worms is the lack of a gut within the body. It is probable that nutrients are absorbed directly from the environment as organic molecules. The large surface-area-to-volume ratio in these animals favors such uptake of nutrients.

The sexes are separate and the sperms are released in packets (spermatophores) which are trapped by the tentacles of the females. The packets gradually disintegrate, and the freed sperms fertilize the eggs, which develop into larvae while still inside the female tube. When large enough they move out and settle nearby.

There has been much discussion about the evolutionary position of the beard worms. It seems likely that the possession of a coelom and the segmentation of the opisthosoma places them near the annelid worms.

Arrow worms are small inconspicuous marine animals that were unknown until collected in 1829 after the invention of the towed plankton net which strained minute life from the sea. Their torpedo-shaped bodies, combined with their ability to dart rapidly forward through the water for short distances, justify the name arrow worms. They are transparent and almost all are colorless. All known species are similar in appearance and all live in the plankton,

except *Spadella*, which lives in the sea floor.

The long narrow body carries a head at the front which bears curved hooks of chitin. These serve to grasp prey and to maneuver it into the mouth. Small planktonic animals form the diet, especially crustaceans such as copepods, but occasionally slightly larger food such as other arrow worms and fish larvae may be taken. The mouth carries small teeth which assist in the act of swallowing. A pair of small simple eyes is also borne on the head. These may or may not be pigmented and serve mainly as light detectors rather than as organs that actually form an image.

The trunk carries two pairs of lateral fins and the body terminates in a fish-like tail segment. The gut runs straight through from mouth to anus and has two small pouches or diverticula at the front.

Arrow worms are hermaphrodite and their sex organs are relatively large for the overall body size. The ovaries lie in the trunk region and the testes and seminal vesicles lie toward the tail separated by a partition. The form of the seminal vesicles, where sperm is stored, is used in identifying the various species. The breeding cycle varies in length according to distribution. It is completed rapidly in tropical waters, in about six weeks in temperate latitudes, and it may take up to two years in polar regions. Fertilization takes place internally and fertilized eggs are released into the sea where they develop as larvae before maturing into adults. The larvae resemble small adults. In the genus *Eukrohnia* the larvae are brooded in special pouches.

The outside of the body is equipped with small sensory tufts which respond to vibrations in the water such as are generated by certain suitable prey species. This facility allows the arrow worms to detect prey and to successfully engulf them. If they themselves suffer an accident, for example in an attack by a predator, lost parts, including the head, can be regenerated.

The evolutionary affinities of the arrow worms remain unclear despite various attempts to show relationships with other phyla. Today they are generally supposed to be without close affinities.   AC

ACORN WORMS

Phylum: Hemichordata
About 90 species in 2 classes.
Distribution: marine; acorn worms worldwide, pterobranchs mainly European and N American waters.
Fossil record: acorn worms, none; pterobranchs possibly rare from Ordovician (500–440 million years ago) to present.
Size: acorn worms 0.8in to 8.2ft (2cm–2.5m); pterobranchs as colonies up to 33ft (10m) in diameter with individuals up to 0.6in (1.4cm) although often less.

Features: essentially worm-like with a coelom developed along deuterostome lines; may be solitary (acorn worms) or colonial (pterobranchs) with a body divided into three zones: proboscis, collar and trunk; nervous system quite well developed; blood circulatory system present; with or without gill slits and tentacles; nephridia lacking; sexes separate.

Acorn worms
Class: Enteropneusta
One order containing, for example, the genera *Balanoglossus, Glossabalanus, Saccoglossus.*

Pterobranchs
Class: Pterobranchia
Two orders containing, for example, the genera *Cephalodiscus, Rhabdopleura.*

HEMICHORDATES may be described as a minor phylum of marine invertebrates because they are not abundant and relatively few species are known. For the zoologist, however, they present a fascinating group of animals since they display a number of features which indicate similarities to the chordates (lancelets, fish, mammals etc), a group with which they were once classified. These characters include gill slits and a nerve cord situated on the upper (dorsal) side of the body. In some species the nerve cord is hollow and like those of vertebrates (fish, mammals etc). At no stage of their development, however, does a rudimentary backbone (notochord) appear, and they are excluded from the phylum Chordata on this technicality. Hemichordates are divided into the two classes Enteropneusta (acorn worms) and Pterobranchia, which have no common name. The pterobranchs are regarded as being more primitive than the acorn worms and some of these lack gill slits and have solid nerve cords. Hemichordates also show similarities with the echinoderms. The larva is a tornaria, which has features in common with certain echinoderm larvae, for example the asteroid bipinnaria. Still more interesting, the pterobranchs also,

with their appendages clothed with tentacles, show similarities to the sea mats (Bryozoa), horseshoe worms (Phoronida) and lampshells (Brachiopoda).

The worm-like acorn worms are the only hemichordates likely to be encountered other than by a scientist, and then only rarely. They grow to 8.2ft (2.5m) in length but are often smaller. Their bodies are made up of three sections, the proboscis at the front, the collar and the trunk at the rear which forms the bulk of the body. It is the way in which the proboscis joins the collar, resembling an acorn sitting in its cup, which gives the group its common name. The proboscis is a small conical structure connected to the collar by a short stalk. The collar itself is cylindrical and runs forward to ensheath the proboscis stalk. The collar bears the mouth on its underside. The trunk makes up most of the body length and at the front end this contains a row of gill slits on each side. A ridge runs down the middle of the back of the trunk. The reproductive organs are borne outside the gill slits; in some forms the trunk is extended as wing-like genital flaps to contain them.

The body cavity (coelom) is present in all three parts of the body. There is a single

▶ **Like a monster from outer space** OPPOSITE, the front region of the acorn worm *Balanoglossus australiensis*, a species found in waters around Australia living in sand under stones.

▶▼ **Body plans of hemichordates.** (1) A colony of individuals of the pterobranch genus *Rhabdopleura*, showing individuals within tubes that are connected by a stolon. (2) Close-up of the head of *Rhabdopleura* protruding from its tube. (3) General body form of the acorn worm *Saccoglossus*. (4) Front end of the acorn worm *Protoglossus*, showing water currents carrying food particles. (5) Burrow system of the acorn worm *Balanoglossus*.

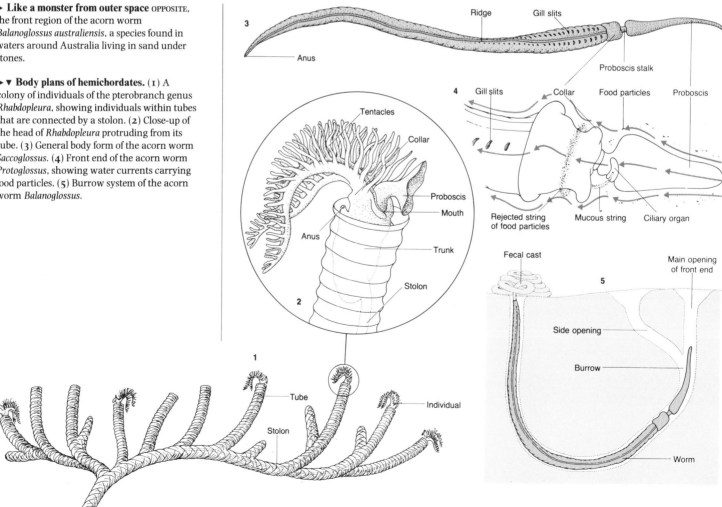

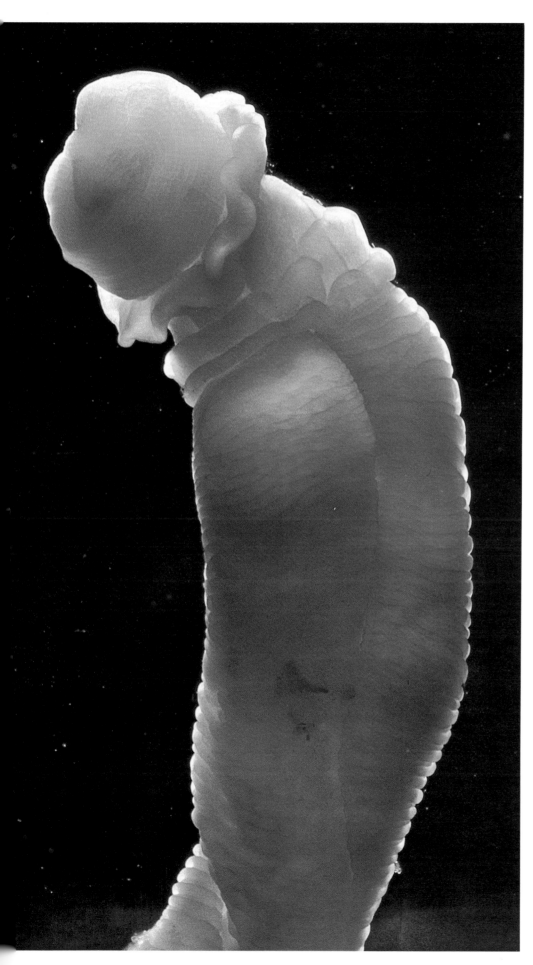

cavity in the proboscis, a pair of cavities in the collar and a pair in the trunk.

Acorn worms live in shallow water and some species may be found burrowing in sandy and muddy shores where they may be recognized by their characteristic coiled fecal casts. The burrows may be more or less permanent. Other species can live under stones and pebbles.

Most of the burrowing acorn worms feed on the organic material in the sand or mud in which they live, simply by ingesting the sediment as an earthworm ingests soil. Some feed by trapping suspended plankton and particles of detritus in the mucus covering on the proboscis. Cilia then pass these particles back to the mouth and the collar can play a role in rejecting unsuitable particles.

The gut runs from the mouth on the collar, via the pharynx of gill slits in the front trunk, to the rear of the trunk where the anus is situated. There is a long, thin blind-ended branch (diverticulum) running from the gut near the mouth into the proboscis, which once was mistaken for a notochord (hence the earlier taxonomic association with the chordates).

The gill slits were originally probably feeding mechanisms, which have subsequently become involved in gas exchange. The cilia they bear pump in water through the mouth and out through the gills.

The nervous system of the acorn worms is fairly primitive by comparison with the chordates, and of course the animals are not highly active. The sexes are separate and the eggs are fertilized in the sea. Initial development resembles that seen in the echinoderms. In some species it proceeds to a tornaria larva which lives in the plankton before it metamorphoses. Other types develop directly, with a juvenile worm appearing.

The pterobranchs are very different, although their bodies still show the three zones. The proboscis is smaller and shield- or plate-like, while the collar is extensively developed with its outgrowths of tentacles. According to the group, there may be two backward-curving arms bearing tentacles (as in *Rhabdopleura* species) or between five and nine (as in *Cephalodiscus*). It is thought that the tentacles function in food gathering.

The gill slits are few and inconspicuous and none is present in *Rhabdopleura*. In this group the gut is U-shaped with an anus opening on the top side of the collar. Again the sexes are separate and many individuals may live grouped together.   AC

ARTHROPODS

Crustaceans
Phylum: Crustacea

Chelicerates
Phylum: Chelicerata

Uniramians
Phylum: Uniramia
All 3 subphyla terrestrial

Water bears
Phylum: Tardigrada

Tongue worms
Phylum: Pentastomida

JOINTED-LIMBED animals or arthropods (*Greek arthron*, joint; *pous, podos*, foot) include the commonest animals in the world today—the insects—as well as such other familiar creatures as the land-dwelling centipedes and spiders and the aquatic crustaceans, including lobsters and crabs.

The most obvious characteristics of an arthropod are the paired jointed limbs or other appendages along the body, and the hard outer covering (exoskeleton) containing chitin which makes their presence possible. Chitin is made up of long fibrous molecules with loose similarities to plant cellulose, and can be formed into many shapes, being truly plastic. For growth to take place, the skeleton must be shed, so arthropods grow in a series of stages of increasing size, the new cuticle hardening after being expanded to the new size.

Arthropods are typically mobile, show bilateral symmetry (right- and left-hand sides of the body being counterparts) and have a distinct head. As in annelid worms, the body is made up of a series of segments, each of which in primitive forms resembles the next and bears a pair of appendages. However, the segments are usually grouped into larger, efficient functional units such as the head, thorax and abdomen of insects and crustaceans, or the prosoma and opisthosoma of spiders. The brain of arthropods consists of dorsal (back, or upper) and ventral (underside, or lower) parts connected by a ring of nervous tissue round the esophagus, and, unless otherwise modified, leads to a double nerve cord on the underside with segmental swellings (ganglia). The body organs are bathed in blood, for the blood system is open, comprising blood-filled spaces with few arteries and veins. Arthropods do have a coelom (the extensive body cavity of earthworms and of ourselves), but it is usually reduced to a number of blind sacs leading via tubules to the outside, and is used for regulation of osmotic pressures within the body and/or for excretion.

Until recently, these and other shared characteristics were believed to indicate that the arthropods should be considered as a single, closely related, evolutionary group— a monophyletic phylum. However, detailed work, particularly in the fields of comparative functional morphology by Dr Sidnie Manton and of comparative embryology by Professor D. T. Anderson, has now shown that there are at least three distinct evolutionary lineages among living arthropods: a chitinous exoskeleton and segmental jointed appendages (along with other common arthropod features) have been evolved together more than once, from unrelated ancestors. Thus compound eyes and biting jaws, apparently so alike in insects and crustaceans, have been evolved separately. Such superficial similarities are to be expected: the same functions are carried out in the two groups, and only certain structural features are capable of fulfilling them.

Dr Manton's comparative study of form and function centered on jaw mechanisms and their associated head structure, and on the nature of body limbs and the locomotory habit in many arthropods. She concluded that three independent living arthropod phyla should be differentiated—the Crustacea, the Chelicerata (including the terrestrial arachnids such as spiders, scorpions and their marine relatives the horseshoe or king crabs and sea spiders) and the Uniramia, which includes the insects. Evidence drawn from the developmental patterns of arthropod embryos, using fate maps to illustrate which original embryonic cells give rise to which adult organ systems, gives strong support to Manton's thesis of three independent living arthropod phyla. The extinct trilobites are also clearly arthropods, and although they do not fall into any of the living arthropod groups, they do show some similarities to the chelicerates. Indeed, trilobites, crustaceans and chelicerates may be more closely related to each other than to uniramians.

Crustaceans are enormously successful in aquatic, particularly marine, environments.

▲ **Always hungry,** the Shore or Green crab investigates a possible food source (here a periwinkle) using its two pairs of antennae and its pincers or chelae. This almost cosmopolitan crab is abundant on rocky shores and in estuaries from Brazil to North America, Europe, North Africa, Sri Lanka, Australia and Hawaii. It is most active at night and at high tide, moving upshore with the tide.

Most living chelicerates are terrestrial arachnids, but the origin of these forms lay in the sea. Marine horseshoe crabs are indeed living fossils, providing a clue as to the nature of the marine ancestors of scorpions and spiders, and are themselves closely related to a large group of fossil chelicerates—the eurypterids or sea scorpions. Sea spiders (Pycnogonida) may well be more recent marine relatives of the arachnids.

Two other living groups of aquatic organisms, the water bears and the tongue worms, show arthropod affinities, particularly the possession of a cuticle (albeit thin) and structures describable as limbs. Various authorities have considered that water bears (tardigrades) should be considered as uniramians and that tongue worms may have crustacean affinities.

Easily the most numerous of arthropods (and of all animals) are the one million or more terrestrial species of Uniramians. They are so called because their limbs consist of a single branch, as opposed to the originally two-branched limbs of crustaceans and chelicerates, although most of the latter have subsequently evolved one-branched limbs as in spiders. In addition to the winged insects (Pterygota), uniramians include the worm-like velvet worms (genus *Peripatus*), those close relatives of the winged insects with six legs, such as the springtails (Collembola) and silverfish (Thysanura), and those multi-legged forms (myriapods), the millipedes and centipedes.

In spite of their names, millipedes do not have as many as a thousand legs nor do centipedes quite have a hundred. Millipedes are typically slow-moving herbivores living under stones or wood. They appear to have two pairs of appendages per segment but each "segment" is strictly formed by fusion of two original ones. Centipedes are typically faster-moving, flattened carnivores with poison claws delivering death to their invertebrate prey.

PSR

CRUSTACEANS

Phylum: Crustacea

About 39,000 species in 10 classes.
Distribution: worldwide, primarily aquatic, in seas, some in fresh water, woodlice on land.
Fossil record: crustaceans appear in the Lower Cambrian, over 530 million years ago.
Size: from microscopic (0.006in/0.15mm) parasites to goose barnacles 30in (75cm) and lobsters 24in (60cm) long.

Features: 2 pairs of antennae; typically segmented body covered by plates of chitinous cuticle subject to molting for growth; body divided into head, thorax and abdomen; head and front of thorax may fuse to form cephalothorax often covered by shield-like carapace; segments from 20 in decapods to 40 in some smaller species; compound eyes; paired appendages on each segment typically include the antennae, main and 2 accessory pairs of jaws (mandibles, maxillules, maxillae), and typically 2-branched limbs on thorax (called pereiopods) and abdomen (generally called pleopods) usually jointed, variously adapted for feeding, swimming, walking, burrowing, respiration, reproduction, defense (pincers); breathing typically by gills or through body surface; body organs in blood-filled hemocoel; double nerve cord on underside; foregut, midgut, hindgut; sexes typically separate; development of young via series of larval stages.

IN addition to many familiar animals, such as crabs and lobsters, the phylum Crustacea includes myriad little-known smaller relatives, some of which are major components of plankton. Crustaceans are clearly arthropods, with their tough, chitin-rich exoskeletons, which require molting for growth to take place, and their paired jointed appendages, or limbs, arranged on segments down the body. They are primarily aquatic, mostly marine, but include fresh-water representatives. One group, the woodlice, has successfully colonized land.

The numerous segments of crustaceans are usually grouped into three functional units (tagmata)—the head, thorax and abdomen. The head is made up of six segments fused with a presegmental region (the acron) while the thorax and abdomen (which often ends in a non-segmental telson) vary in number of segments. Some of the smaller crustaceans contain up to 40 segments, whereas the decapods (shrimps, crabs, lobsters), the biggest crustaceans, have 20—six in the head, eight in the thorax and six in the abdomen. Some thoracic appendages may fuse to the back of the head, forming a cephalothorax, and a shield-like carapace may cover the head and part or all of the thorax: crustacean evolution is the story of reduction in the number of segments and their grouping into tagmata, with consequent specialization of the appendages for specialized roles. Although the fossil record goes back well over 500 million years, it tells us little about the origin of crustaceans (see pp110–117) for the basis of deductions concerning evolutionary changes within the crustaceans).

The "typical" crustacean head bears five pairs of appendages. There are none on the first segment, but the following two segments both bear a pair of antennae—the first antennae (antennules) and the second antennae (antennae). All crustaceans have two pairs of antennae. The antennae have probably shifted in evolution, for they now lie in front of the mouth, which may have been terminal in a pre-crustacean ancestor. They are typically sensory, although in some crustaceans they are employed in locomotion or even used by a male to clasp the female in reproduction.

Behind the mouth is the fourth segment bearing the mandibles, the main jaws developed from jaw-like extensions (gnathobases) at the base of the appendages, while the remainder of the limb has been lost, or reduced to a sensory palp. The fifth and sixth head segments bear the first maxillae (maxillules) and second, or true, maxillae respectively; these accessory jaws assist

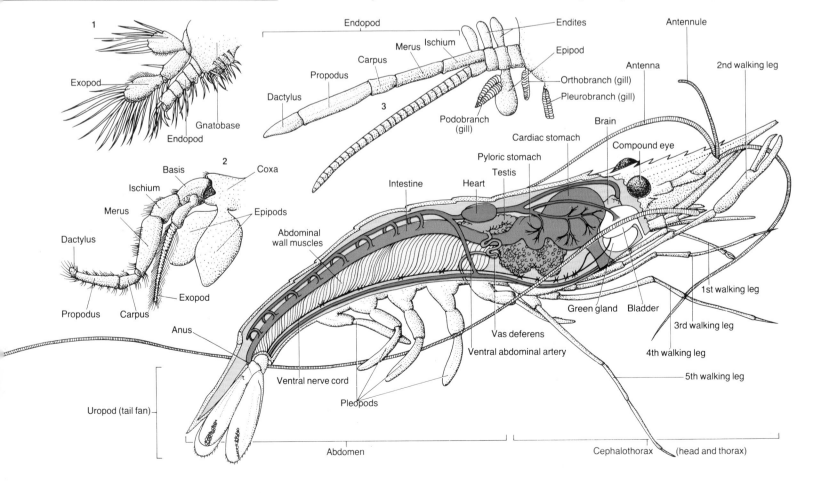

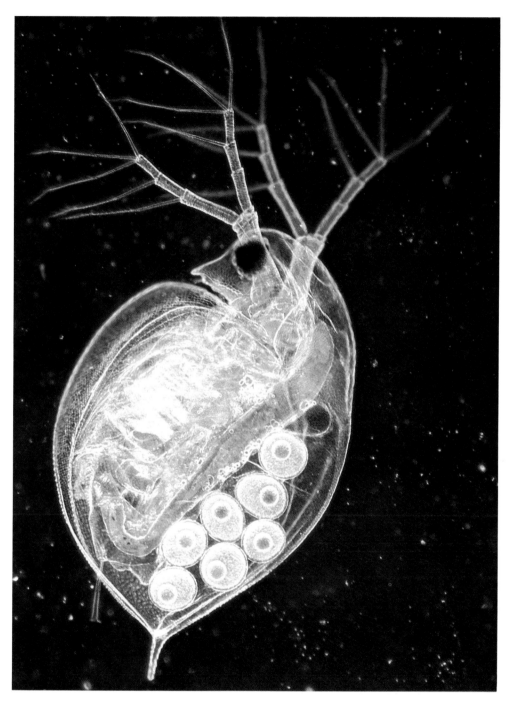

▲ Using its head to swim: the powerful two-branched antennae of the tiny freshwater flea *Daphnia* provide the jerking power-stroke to lift the animal through the water. This female bears seven embryos in the brood pouch between her abdomen and carapace.

◄ Body plan of a crustacean, showing the main typical external features. Illustrated is *Palaemon*, a genus of free-swimming prawns of the order Decapoda.

Crustacean limbs range from the primitive and leaf-like "phyllopods" to advanced walking legs as in lobsters. Shown here are thoracic appendages of (1) *Hutchinsoniella* (class Cephalocarida); (2) a malacostracan of the superorder Syncarida; (3) a decapod malacostracan (superorder Eucarida).

mastication and are similarly derived by partial or complete reduction of the limb to a gnathobase.

The ancestors of crustaceans were probably small marine organisms living on the sea bottom but able to swim, with a series of similar appendages down the length of a body not divided into thorax and abdomen; all appendages would have been used in locomotion, feeding and respiration. The limbs of early crustaceans were two-branched (biramous), with an endopod (inner limb) and exopod (outer limb) branching from the base (protopodite). The ancestral limb would have had two flat leaf-like lobes, as in several living crustaceans of

the order Branchiopoda (eg fairy shrimps and brine shrimps). Such limbs are usually moved in rhythm to create swimming and feeding currents, food being passed along the underside of the body to the mouth—so promoting the evolution of jaws from limb gnathobases. Extensions from the outer side of the limb bases often act as respiratory surfaces.

One major line of crustacean evolution has produced large walking animals. The two-branched swimming and filtering leg has evolved into an apparently one-branched (uniramous) walking leg (stenopodium)—strictly the endopod alone, the exopod being reduced and lost during larval development or previously in evolution. Such cylindrical legs have a reduced surface area and are not suitable as respiratory surfaces, which are particularly necessary in large crustaceans. Exites or extensions of the body wall at the base of the legs are therefore used as gills. The decapods (shrimps, crabs, lobsters) also show increased development of the head (cephalization). The first three pairs of thoracic limbs are adapted as accessory mouthparts (maxillipeds) and a carapace shields the cephalothorax.

Crustaceans typically have separate sexes and the fertilized egg as it divides shows a characteristic modified spiral cleavage pattern. Crustaceans usually develop through a series of larval stages of increasing size and numbers of segments with their associated limbs. The simplest larval stage is the nauplius larva, with three pairs of appendages—the first and second antennae and the mandibles. It occurs in many living crustaceans, often as a pelagic dispersal stage that swims and suspension-feeds with the three pairs of appendages. These limbs therefore have different functions in larva and adult, usually relinquishing their larval roles to other appendages which develop further down the body. Nauplius larvae are followed by a variety of larger larvae, according to the type of crustacean, although many crustaceans bypass the nauplius equivalent by developing within the egg.

The chitin-rich exoskeleton of crustaceans is divided into a series of plates which may be hardened by calcification or by tanning. The body organs lie in a blood-filled space—the hemocoel, and the blood often contains hemocyanin, a copper-based respiratory pigment which bears oxygen and is equivalent to hemoglobin in vertebrates. The coelom, the major body cavity of many animals, is restricted to the inner

cavity of the coxal glands—paired excretory and osmoregulatory organs each consisting of an end sac and a tubule that opens out at the base of the second antenna (antennal gland) or the second maxilla (maxillary gland).

The alimentary canal consists of a cuticle-lined foregut and hindgut, with an intermediate midgut giving rise to blind pockets (ceca), perhaps modified as a hepatopancreas combining the functions of digestion, absorption and storage. The nervous system consists of a double ventral nerve cord, primitively with concentrations of the cord (ganglia) in each segment, but often further concentrated into a few large ganglia.

The following account covers the more important crustacean groups. For a complete list of classes within the phylum Crustacea, see overleaf.

Cephalocarids

Cephalocarids, first described only in 1955, are often considered to be the most primitive of living crustaceans. The nine species of the class Cephalocarida so far discovered are all minute inhabitants of marine sediments, feeding on bottom detritus. The body, under 0.2in (4mm) in length, is not divided into thorax and abdomen, and the leaf-shaped body appendages are very similar to each other and indeed to the second maxillae. Many experts believe the cephalocarid limb to represent an ancestral type from which the limbs of other living crustaceans have evolved.

Fairy shrimps and water fleas

Fairy shrimps and water fleas are among the branchiopods, small, mostly freshwater, crustaceans. They have leaf-like trunk appendages used for swimming and filter-feeding. Flattened extensions (epipodites) from the first segment (coxa) of these limbs act as gills, giving the class the name of "gill legs" (Branchio-poda).

Fairy shrimps (order Anostraca) live in temporary freshwater pools and springs, (some, the brine shrimps, in salt lakes), typi-

cally in the absence of fish. Unlike other branchiopods they lack a carapace (hence An-ostraca). They have elongated bodies with 20 or more trunk segments, many bearing appendages of the one type. They are usually about 0.4in (1cm) long, but some giants reach 4in (10cm). Fairy shrimps swim upside down, beating the trunk appendages in rhythm, simultaneously filtering small particles with fine slender spines (setae) on the legs. Collected food particles are then passed along a groove on the underside to the mouth.

When mating, a male fairy shrimp clasps the female with its large second antennae. The female lays her eggs into a brood sac. The eggs on release are extremely resistant to drying out (desiccation). Some eggs will hatch when wetted, but others require more than one inundation—a successful evolutionary strategy, ensuring that populations are not wiped out when insufficient rain falls to maintain the pool long enough for completion of the fairy shrimp life cycle. Fairy shrimps hatch as nauplii and grow to maturity in as little as one week.

The brine shrimps are found in salt lakes, the brackish nature of which eliminates possible predators. The eggs similarly resist drying out and are sold as aquarium food. *Artemia salina* has a remarkable resistance to salt, surviving immersion in saturated salt solutions. The gills absorb or excrete ions as appropriate and this brine shrimp can produce a concentrated urine from the maxillary glands.

Water fleas (order Cladocera) are laterally compressed with a carapace enclosing the trunk but not the head, which projects on the underside as a beak. Overall length is just 0.04–0.2in (1–5mm). The powerful second antennae are used for swimming. The trunk is usually reduced to about five segments, of which two may bear filtering appendages. Most water fleas, including *Daphnia* species, live in freshwater and filter small particles, but some marine cladocerans are carnivorous.

Water fleas brood their eggs in a dorsal

▲► **Representative species of planktonic crustaceans.** (1) A species of *Sagitta*, a genus of arrow worms which feed on copepods (0.4–1.6in, 1–4cm). (2) A species of *Oikopleura*, a genus of larvaceans, dwelling inside a gelatinous protective "house" (1.2in, 3cm). (3) A free-living copepod. It swims by using its well-developed antennae. (4) A phyllosoma larva of a crawfish. This delicate planktonic larva develops into the massive bottom-dwelling crawfish shown on p96. (5) A megalopa, the late larval stage of crabs which has head, thoracic and abdominal appendages. (6) A caprellid, a minute bottom-dwelling crustacean. These are often less than 0.04in (1mm) long and often live in association with other invertebrates or plants. (7) *Salpa fusiformis*, a planktonic relative of the sea squirts. Individuals form chains of varying lengths and drift in surface waters. (8) A zoea larva, ie the early larval stage of a crab. It has a full complement of head appendages but only the first two pairs of thoracic appendages. After several molts it develops into a megalopa. (9) *Pycnogonum littorale*, a sea spider which lives on the lower shore and in shallow water under stones etc in NW Europe (0.8in, 2cm).

chamber and asexual reproduction (parthenogenesis) is common. Populations may consist only of females reproducing parthenogenetically for several generations until a temperature change or limitation of food supply induces the production of males. The brood chamber, which now encloses fertilized eggs, may be cast off at the next molt and can withstand drying, freezing and even passage through the guts of vertebrates. Some *Daphnia* species show cyclomorphosis—cyclic seasonal changes in morphology, often involving a change in head shape.

In the clam shrimps (order Conchostraca), most of which are some 0.4in (1cm) in length, the remarkably clam-like bivalve carapace encloses the whole body, but in tadpole shrimps (order Notostraca) part of the trunk extends behind the large shield-like carapace and the overall body length may reach 2in (5cm). Either may be found with fairy shrimps in temporary rain pools.

Copepods

The class Copepoda includes over 180 families of mostly minute sea- and fresh-water inhabitants which provide a major source of food for fish, mollusks, crustaceans and other aquatic animals. As dominant members of the marine plankton, copepods of the order Calanoida may be among the most abundant animals in the world. On the other hand, members of the order Harpacticoida are common inhabitants of usually marine sediments, and Cyclopoida species may be either planktonic or bottom dwelling (benthic) in the sea or freshwater. Some copepods are parasitic, infesting other invertebrates, fish (often as fish lice on the gills) and even whales. Some of these attain 12in (30cm) in length.

Free-living copepods are small and capable of rapid population turnover. The large first antennae may be used for a quick escape, but more commonly they act as parachutes against sinking, while the thoracic limbs are the major swimming organs.

Many calanoid copepods can filter feed on phytoplankton, using the feathery second maxillae to sieve a current driven by the second antennae which beat at about 1,000 strokes per minute. The setae on the second maxillae are adapted to trap a particular size of microscopic plant organisms (phytoplankton) and the first pair of thoracic limbs (maxillipeds) scrape off the filtered material.

In fact calanoid copepods cannot survive in the marine planktonic ecosystem by filtering alone. Filter feeding is an energy-

▲ **Conspicuous in many food webs,** copepods occur even in the ocean depths. *Megacalanus princeps* is a large, pigmented species found at depths of 1,640–3,280ft (500–1,000m).

Feeding largely on microscopic plants, especially during the spring bloom, planktonic copepods are themselves a major food item for fishes, mollusks and other sea creatures.

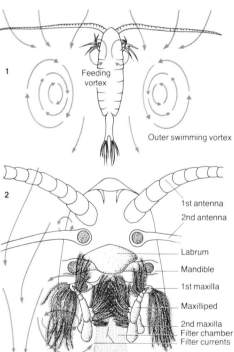

◀ **Techniques of filter feeding** in a copepod (*Calanus* species): (1) swimming and feeding currents created by the feathery second antennae; (2) detail of filter currents and filter apparatus.

1 Feeding vortex

Outer swimming vortex

2
1st antenna
2nd antenna
Labrum
Mandible
1st maxilla
Maxilliped
2nd maxilla
Filter chamber
Filter currents

▶ **Opportunistic goose barnacles** OVERLEAF hang down below floating pumice stone in the warm waters of the Great Barrier Reef off Australia. From the shell plates, six thoracic cirri protrude to filter out food. The species illustrated is *Lepas anserifera*.

sapping process that requires the driving of large volumes of water through a fine sieve. There must therefore be a minimum level of phytoplankton in the sea simply to repay the cost of filtering. Phytoplankton populations in temperate oceans only reach such concentrations for brief periods in the year—in the North Atlantic, during the spring bloom and perhaps again in the fall. For much of the year, therefore, the calanoid copepod feeds raptorially—seizing large

Class Cephalocarida

One order, 2 families, 4 genera, 9 species (eg *Hutchinsoniella macracantha*). Primitive bottom dwellers feeding on deposited detritus.

Branchiopods
Class Branchiopoda

Mostly small freshwater filter feeders. Comprises the **tadpole shrimps** (order Notostraca), 11 species in 1 family, eg *Triops* species; **clam shrimps** (order Conchostraca), 180 species in 5 families, eg *Cyzicus* species; **water fleas** (order Cladocera), 450 species in 11 families, eg *Daphnia* species; and **fairy shrimps** (order Anostraca) 180 species in 7 families, eg **brine shrimps**, *Artemia* species.

Class Remipedia
One species (*Speleonectes lucayensis*), 0.4in (5mm) long, a blind swimmer in marine caves.

Class Tantulocarida
Four species (eg *Basipodella harpacticola*) in 1 family, microscopic 0.006in (0.15mm) ectoparasites on deepsea bottom-dwelling crustaceans.

Class Mystacocarida

Nine species in 1 family, minute (0.02in/0.5mm) in sediments of seabed feeding on detritus. Eg *Derocheilocaris typicus*.

Branchiurans
Class Branchiura

One order (150 species) of blood-sucking **fish lice** some 0.3in (7mm) long, ectoparastic on marine or freshwater fish. Eg *Argulus foliaceus*.

Copepods
Class Copepoda

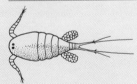

Over 8,400 species in 181 families and 8 orders of mostly minute components of plankton and sediment fauna. Includes: order Calanoida (2,300 species, eg *Calanus tinmarchicus*, in 40 families); order Harpacticoida (2,800 species, eg *Harpacticus* species, in 34 families); order Cyclopoida (450 species, eg *Cyclops* species, in 14 families); and the orders Poecilostomatoida (1,320 species, eg *Ergasilus sieboldi*, in 41 families) and Siphonostomatoida (1,430 species, eg *Caligus* species, in 44 families) of **fish lice**.

Barnacles
Class Cirripedia

Four orders comprising 1,025 species, adults parasitic or permanently attached to substrate, most with long, curved filter-feeding "legs." Comprising order Ascothoracica (45 species, eg *Ascothorax* species, in 4 families), very small parasites on echinoderms and soft corals; order Acrothoracica (50 species, eg *Trypetesa* species, in 3 families), small (0.16in/0.4mm) filter feeders boring into calcareous substrates (eg shells); order Thoracica (700 species in 17 families), filter-feeding barnacles on rocks etc, up to 30in (75cm) long, including **goose barnacles** (*Lepas* species), **acorn** or **rock barnacles** (*Balanus* species) and **whale barnacles** (*Conchoderma* species); and order Rhizocephala (230 species, eg *Sacculina carcini*, in 6 families), internal parasites of decapods (lobsters, shrimps, crabs).

Mussel shrimps and seed shrimps
Class Ostracoda

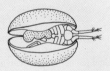

Five orders comprising 54 families and 5,650 species. Small, 0.04–0.12in (1–3mm), mostly bottom dwelling, with two-valved carapace. Eg *Cypris* species.

Malacostracans
Class Malacostraca

Fourteen orders, 359 families, 23,000 species.

Subclass Phyllocarida
Order Leptostraca
One family containing 25 species of pelagic or bottom-dwelling feeders on detritus. Eg *Nebalia bipes*.

Subclass Hoplocarida
Order Stomatopoda
Twelve families containing 350 species of bottom-dwelling carnivorous **mantis shrimps** (eg *Squilla empusa*).

Subclass Eumalacostraca
Four superorders.

Superorder Syncarida
Comprises the orders Anaspidacea (15 species, eg *Anaspides tasmaniae*, in 4 families), S Hemisphere freshwater bottom-dwelling detritus feeders up to 2in (5cm) long; and Bathynellacea (100 species, eg *Parabathynella neotropica*, in 3 families), inhabiting freshwater sediments feeding on detritus, blind, 0.02–0.04in (0.5–1.2mm) long.

Superorder Pancarida
Order Thermosbaenacea, 1 family of 9 species, eg *Thermosbaena mirabilis*, of sediment-dwellers in marine and fresh water, and in hot-spring ground-water, about 0.12in (3mm) long.

Superorder Peracarida
Six orders, including: the **opossum shrimps** (order Mysidacea, 780 species, eg *Neomysis integer*, in 6 families), free-swimming and bottom-dwelling filter feeders 0.4–1.2in (1–3cm) long; **sea slaters, woodlice, pill bugs** and **sow bugs** (order Isopoda, 4,000 species in 100 families), eg *Armadillidium, Oniscus, Porcellio*, marine, freshwater and terrestrial, typically omnivorous crawlers but some parasites (eg *Bopyrus* species) and wood-boring **gribbles** (*Limnoria* species), 0.2–0.6in (0.5–1.5cm) long; **sand hoppers** and **beach fleas** (order Amphipoda, 6,000 species in 96 families), mostly marine bottom-dwelling scavengers, 0.04–0.06in (0.1–1.5cm) long, but also **skeleton shrimps** (*Caprella* species), **whale lice** (*Cyamus* species) and the "**Freshwater shrimp**" (*Gammarus pulex*).

Superoder Eucarida
Three orders: **krill** (order Euphausiacea), 85 species, eg **Whale krill** (*Euphausia superba*), in 2 families of marine pelagic filter feeders 0.2–2in (0.5–5cm) long; order Amphionidacea, 1 marine free-swimming species, *Amphionides reynaudii*, to 1.2in (3cm) long; and the **decapods** (shrimps, lobsters, crabs) (order Decapoda), 10,000 species in 105 families.

See pp98–99 for Table of Decapods.

prey items including other members of the zooplankton. During the rich spring "bloom," filtering becomes worthwhile and the copepods can pass through several generations, multiplying very quickly, before returning to low metabolic tickover, often using fat reserves for the rest of the year. The number of copepod generations per year is related to the time period of the phytoplankton bloom.

Copepods pair, rather than copulate directly, the male transferring a packet of sperm (spermatophore) to the female, perhaps in response to a "chemical message" (pheromone) from her. The eggs are fertilized as they are laid into egg sacs carried by the female. Eggs hatch as nauplii and pass through further nauplius and characteristic copepodite stages before adulthood.

It is the larval stages that ensure the dispersal of parasitic copepods. Copepods that are parasitic on the exterior of their host (exoparasites) may show little anatomical modification, but endoparasites often consist of litle more than an attachment organ and a grossly enlarged genital segment with large attendant egg sacs.

Barnacles

Barnacles are sedentary marine crustaceans, permanently attached to the substrate. For protection barnacles have carried calcification of the cuticle to an extreme and have a shell resembling that of a mollusk. The shell is a derivative of the cuticle of the barnacle head and encloses the rest of the body. Stalked goose barnacles, which commonly hang down from floating logs, are more primitive than the acorn barnacles that abut directly against the rock and dominate temperate shores. (For barnacles' economic importance, see pp106–107.)

Barnacles (class Cirripedia) feed with six pairs of thoracic legs (cirri) which can protrude through the shell plates to filter food suspended in the seawater. Goose barnacles trap animal prey, but most acorn barnacles have also evolved the ability to filter fine material, including phytoplankton and even bacteria, with the anterior cirri.

Barnacles are hermaphrodites. They usually carry out cross-fertilization between neighbors. Fertilized egg masses are held in the shell until release as first-stage nauplii. Indeed it was J. Vaughan Thompson's observations (1829) of barnacle nauplius larvae of undoubted crustacean pedigree that removed lingering suspicions regarding the possible molluskan nature of barnacles. There are six nauplius stages of increasing size which swim and filter phytoplankton over a period of a month or so, before giving rise to a non-feeding larva—the cypris larva, named for its similarity to the mussel shrimp genus *Cypris*. This is the settlement stage of the life cycle, able to drift and swim in the plankton before alighting, and choosing a settlement site in response to environmental

Barnacles

Barnacles are the most successful marine fouling organisms and more than 20 percent of all known species have been recorded living on man-made objects, including ships, buoys and cables. Goose barnacles (*Lepas* species) attach themselves in enormous clumps to slow-moving sailing ships. They have been replaced today by acorn (or rock) barnacles (*Balanus* species) on motor-powered vessels, although stalked whale barnacles (*Conchoderma* species, named for one of their most important living hosts) are also important on very large crude oil carriers traveling between the Gulf and Europe.

The minute planktonic (nauplius) larvae of barnacles, feeding on phytoplankton, are dispersed in the sea and build up fat reserves to support the non-feeding settlement (cyprid) stage through further dispersal, site selection and metamorphosis. The cyprid, a motive pupa according to Darwin in the 1850s, has a low metabolic rate and lasts as long as a month before alighting on a chosen substrate. The cyprid then walks, using sticky secretions on the adhesive disks of the antennules, responding to current strength and direction, light direction, contour, surface roughness and the presence of other barnacles, as it monitors the suitability of a site for future growth. Finally, cypris cement is secreted from paired cement glands down ducts to the adhesive disks so that they become embedded permanently, whereupon the cypris metamorphoses to the juvenile barnacle which feeds and grows, developing its own cement system.

Barnacles slow down vessels, costing fuel. To counter this, antifouling paint is used which releases toxin in sufficient concentration to kill settling larvae or spores. Copper, the most common antifouling agent, is very toxic and must be released at a rate of 25 micrograms per sq in per day to prevent barnacle settlement and growth.

Barnacles accumulate heavy metals, specifically zinc as detoxified granules of zinc phosphate. Barnacles are therefore suitable monitors of zinc availability in the marine environment, their high concentrations being easily measurable. Thames estuary barnacles, for example, may contain the fantastic zinc concentration of 15 percent of their dry weight.

factors which the larva detects by an array of sense organs.

Barnacles colonize a variety of substrates, including living animals such as crabs, turtles, sea snakes and whales—with a moving host the barnacle does not need to use energy in beating its cirri. Some barnacles have evolved to become parasites, which bear little similarity to their free-living relatives, except as larvae. Members of the order Rhizocephala (literally "root-headed") parasitize decapod crustaceans.

Mussel shrimps

Mussel shrimps or ostracods (class Ostracoda) are small bivalved crustaceans widespread in sea- and fresh water. Some are planktonic, but most live near the bottom, plowing through the detritus on which they may feed. Algae are another favored food.

The rounded valves of the carapace completely enclose the body, which consists of a large head and a reduced trunk, usually with only two pairs of thoracic limbs. The antennae are the major locomotory organs. Both pairs may be endowed with long bristles (setae) to aid propulsion when swimming, or the first antennae may be stout for digging, or even hooked for climbing aquatic vegetation.

Malacostracans

Including more than half of all living crustacean species, the class Malacostraca is very important in marine ecology and has also successfully invaded freshwater and land habitats.

Malacostracans are modifications of a shrimp-like body plan. The thorax consists of eight segments bearing limbs, of which up to three of the front pairs are accessory mouthparts or maxillipeds. A carapace typically covers head and thorax, though this may have been lost in some peracarids (see below). Members of the primitive superorder Syncarida also lack a carapace but this absence may be of more ancient ancestry: the anaspidaceans are bottom-dwelling feeders on detritus in Southern Hemisphere freshwaters, and the bathynellaceans are minute, blind detritivores living in freshwater sediments.

Most adult malacostracans are bottom dwellers (benthic), and the single-branched (uniramous) thoracic legs are adapted for walking. Some malacostracans can swim using pleopods, the limbs of the first five abdominal segments. The appendages of the sixth abdominal segment are directed back as uropods to flank the terminal telson and form a tail fan.

The gills of malacostracans are usually situated at the base of the thoracic legs in a chamber formed by the carapace, and are aerated by a current driven forward by a paddle on the second maxillae. Malacostracans typically ingest food in relatively large pieces. The anterior part of the stomach is where the large food particles are masticated before they pass to the pyloric stomach, where small particles are filtered and diverted to the hepatopancreas for digestion and absorption. Remaining large particles pass down the midgut and hindgut to be voided.

The most primitive malacostracans are to be found in the order Leptostraca. They are marine feeders on detritus, with a bivalved carapace, leaf-like thoracic limbs and eight abdominal segments. Mantis shrimps (order Stomatopoda) are marine carnivores which wait at their burrow entrances for unsuspecting prey, including fish, before striking rapidly with the enormous claws on the second thoracic appendages. The small sediment-dwelling crustaceans of the order Thermosbaenacea inhabit thermal springs, fresh and brackish lakes, and coastal ground water.

The vast majority of the remaining malacostracans are grouped within the two superorders Eucarida (see p97) and Peracarida. The peracarids are a superorder of malacostracans containing six orders of which the isopods (woodlice or pill bugs and others) and amphipods (sand hoppers and beach fleas) contain 10,000 species between them. Peracarids hold their fertilized eggs in a brood pouch formed on the underside of the body by extensions from the first segments (coxae) of the thoracic legs, the eggs hatching directly as miniature versions of their parents. The carapace, when present, is not fused to all the thoracic segments. The first thoracic segment is joined to the head.

Opossum shrimps (order Mysidacea), although typically marine, are common also in estuaries, where they may be found swimming in large swarms and may constitute the major food of many fish. They filter small food particles with their two-branched thoracic limbs, and on many species a balancing organ (statocyst) is clearly visible on each inner branch of the pair of uropods.

Their common name comes from the large brood pouch on the underside of the female's thorax.

Woodlice (pill bugs) and other isopods

Woodlice (or pill bugs) are the crustacean

success story on land. They are the most familiar members of the order Isopoda. Like other isopods, they are flattened top-to-bottom (dorsoventrally) and lack a carapace covering the segments. The first pair of thoracic limbs is adapted as a pair of maxillipeds, leaving seven pairs of single-branched thoracic walking legs (pereiopods). The five pairs of abdominal pleopods are adapted as respiratory surfaces. Isopods typically molt in two halves, the exoskeleton being shed in separate front and rear portions.

Most isopods walk on the sea bed, but some swim, using the pleopods, and others may burrow. The wood-boring gribble used to destroy wooden piers along the coasts of the North Atlantic by rasping with its file-like mandibles, a major pest until concrete replaced wood. Isopods are also to be found in freshwater, and sea slaters live at the top of the intertidal zone, indicating the evolutionary route of the better known woodlice to life on land.

The direct development characteristic of reproduction in the superorder Peracarida (see above) avoids the release of planktonic larvae, and is a major adaptive preparation

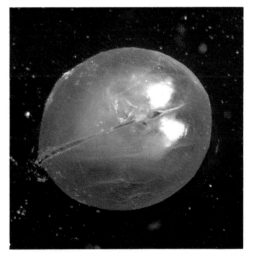

for terrestrial life. Woodlice have behavioral responses, for example to changes in humidity, to avoid desiccatory conditions and often therefore select damp micro-habitats. Members of the pill bug genus *Armadillidium* are able to roll up as the generic name suggests.

Woodlice have adapted their respiratory organs to take up oxygen from air. Members of the genus *Oniscus* show only little change from the ancestral aquatic isopod arrangement of pleopods as gills. Each of the five pairs of pleopods is two-branched and overlaps the one behind. The exopods of the first pair are extensive enough to cover all the remaining pleopods. The innermost fifth pleopods therefore lie in a humid microchamber and the endopods are well supplied with blood to act as the respiratory surface. *Porcellio* and *Armadillidium* species tolerate dryer conditions than can *Oniscus* species and use the outlying exopods of the first pair of pleopods for respiration. The danger of desiccation is reduced, for these exopods have in-tuckings of the cuticle (pseudotracheae) as sites of respiratory exchange.

Most isopods are scavenging omnivores, some tending to a diet of plant matter, especially the woodlice which contain bacteria in the gut to digest cellulose. Of the more carnivorous, *Cirolana* species can be an extensive nuisance to lobster fishermen, devouring bait in lobster pots.

Isopods have also evolved into parasites. Ectoparasitic isopods attach to fish with hooks, and pierce the skin with their mandibles to draw the blood on which they feed. Those isopods (eg *Bopyrus* species) that live in the gill chambers of decapod crustaceans (crabs, shrimps, lobsters) are more highly modified and may cause galls. Some isopods even hyperparasitize rhizocephalan barnacles, themselves parasitic on decapods.

▲ **Hopping for their lives:** Sand hoppers (*Orchestia gammarella*) can use their appendages to jump and thus escape predators, but they are also good swimmers.

▶ **Encased in two shells** ABOVE, the soft body of the mussel shrimp *Gigantocypris* is protected from predators.

◀▶ **Terrestrial wood louse and shoreline sea slater** are related members of the order Isopoda. Common woodlice LEFT (*Armadillidium vulgare*) are familiar land crustaceans able to feed on plant matter and rotting wood. The flattened bodies RIGHT of this male and this female sea slater (*Ligia oceanica*) enable them to shelter in crevices.

Beach fleas and other amphipods

Amphipods (order Amphipoda) are laterally compressed peracarids which often lie on their flattened sides. The most familiar are the beach fleas or sand hoppers found on sandy shores, and the misnamed "Freshwater shrimp" common in European streams. Amphipods are mostly marine. Although they are typically bottom dwellers (benthic), some are free swimming (pelagic).

Like the isopods, amphipods have no carapace, their eyes are stalkless, they have one pair of maxillipeds, seven pairs of single-branched walking legs (pereiopods) and a brooding pouch on the underside. Unlike the isopods, they have gills on the thoracic legs, and on the six abdominal segments there are usually three pairs of pleopods and three pairs of backward-directed uropods.

Most amphipods are scavengers of detritus, able to both creep using the thoracic pereiopods and swim with the abdominal pleopods. Many burrow or construct tubes

and may feed by scraping sand grains or filtering plankton with bristle-covered limbs. Beach fleas, which have a remarkable ability to jump, burrow at the top of sandy shores or live in the strand line. This proximity to dry land has facilitated the evolution of terrestrial amphipods in moist forest litter though to a more limited extent than in isopods.

Amphipods of the family Hyperiidae are pelagic carnivores that live in association with gelatinous organisms, such as jellyfish, on which they prey. *Phronima* species are often reported to construct a house from remains of salp tunicates in which the animal rears its young. Skeleton shrimps (family Caprellidae) are predators of hydroid polyps, and the atypical, dorso-ventrally flattened whale lice are ectoparasites of whales.

Krill and decapods—the eucarids

The planktonic krill and the decapods

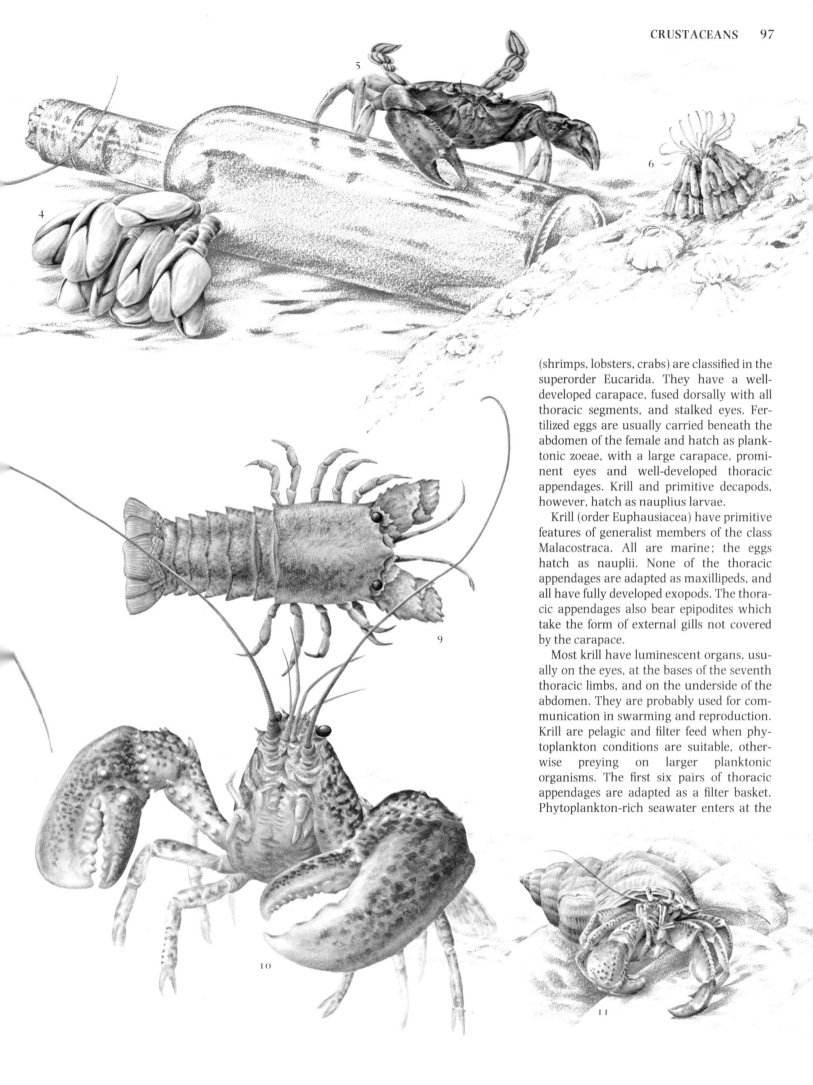

(shrimps, lobsters, crabs) are classified in the superorder Eucarida. They have a well-developed carapace, fused dorsally with all thoracic segments, and stalked eyes. Fertilized eggs are usually carried beneath the abdomen of the female and hatch as planktonic zoeae, with a large carapace, prominent eyes and well-developed thoracic appendages. Krill and primitive decapods, however, hatch as nauplius larvae.

Krill (order Euphausiacea) have primitive features of generalist members of the class Malacostraca. All are marine; the eggs hatch as nauplii. None of the thoracic appendages are adapted as maxillipeds, and all have fully developed exopods. The thoracic appendages also bear epipodites which take the form of external gills not covered by the carapace.

Most krill have luminescent organs, usually on the eyes, at the bases of the seventh thoracic limbs, and on the underside of the abdomen. They are probably used for communication in swarming and reproduction. Krill are pelagic and filter feed when phytoplankton conditions are suitable, otherwise preying on larger planktonic organisms. The first six pairs of thoracic appendages are adapted as a filter basket. Phytoplankton-rich seawater enters at the tips of the legs and is strained as it passes between the leg bases. Whale krill reach about 2in (5cm) long, dominate the zooplankton of the Antarctic Ocean and are the chief food of many baleen whales (see pp18–19).

In decapods, the first three pairs of thoracic appendages are adapted as auxiliary mouthparts (maxillipeds), theoretically leaving five pairs as legs (pereiopods)—hence deca-poda "ten legs." (In fact the first pair of pereiopods is often adapted as claws.) Historically decapods have been divided into swimmers (natantians) and crawlers (reptantians)—essentially the shrimps and prawns on the one hand and the lobsters, crayfish and crabs on the other. More modern divisions rely on morphological characteristics, and decapods are now divided between the suborders Dendrobranchiata and Pleocyemata.

Members of the Dendrobranchiata are all shrimp-like, characterized by their laterally compressed body and many-branched (dendrobranchiate) gills. Their eggs are planktonic and hatch as nauplius larvae. The Pleocyemata have gills which lack secondary branches, being plate-like (lamellate) or thread-like (filamentous) and their eggs are carried on the pleopods of the female before hatching as zoeae.

Prawns and shrimps

The terms shrimp and prawn have no exact zoological definition, and they are often interchangeable.

Prawns, Shrimps, Lobsters, Crabs (Order Decapoda)

Suborder Dendrobranchiata

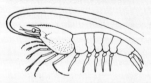

Five families of free-swimmers with many-branched gills and free-floating eggs hatching as nauplius larvae; carnivorous, 0.2–8in (0.5–20cm) long. The family Penaeidae includes **Banana prawn** (*Penaeus merguiensis*), **Brown shrimp** (*P. aztecus*), **Giant tiger prawn** (*P. monodon*), **Green tiger prawn** (*P. semisulcatus*), **Indian prawn** (*P. indicus*), **Kuruma shrimp** (*P. japonicus*), **Pink shrimp** (*P. duorarum*), **White shrimp** (*P. setiferus*), and **Yellow prawn** (*Metapenaeus brevicornis*).

Suborder Pleocyemata

23 families; gills plate-like (lamellate) or thread-like (filamentous), not many-branched; eggs carried by female before hatching as zoeae.

Banded or cleaner shrimps
Infraorder Stenopodidea

One family (Stenopodidae) of bottom-dwelling cleaners of fish, about 2in (5cm) long. Eg *Stenopus* species.

Shrimps, prawns, pistol shrimps
Infraorder Caridea

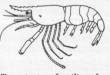

Twenty-two families of marine, brackish and freshwater swimmers and walkers; predatory scavengers, dominant shrimps in N oceans; 0.2–8in (0.5–20cm) long. Includes **Brown shrimp** (*Crangon crangon*) and **Pink shrimp** (*Pandalus montagui*).

Lobsters, freshwater crayfish, scampi
Infraorder Astacidea

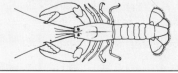

Five families of mostly marine, some freshwater, bottom walkers, hole dwellers; predatory scavengers; up to 2ft (60cm) long. Includes **American lobster** (*Homarus americanus*), **European lobster** (*H. gammarus*) and **Dublin Bay** or **Norway lobster** (*Nephrops norvegicus*).

Spiny and Spanish lobsters
Infraorder Palinura

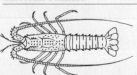

Five families, of marine bottom walkers or hole dwellers; often spiny but lack permanent rostrum (frontal spine) of Astacidea; predatory scavengers up to 2ft (60cm) long. Includes **American spiny lobsters** (*Panulirus argus* and *Panulirus interruptus*) and **European spiny lobster** (*Palinurus elephas*).

◄ **Banded shrimps** (here *Stenopus hispidus*) provide a cleaning service for fish with wounds or parasites. The characteristic coloration may help the "patient" to locate them.

▼ **In a strange procession,** East coast American spiny lobsters migrate annually on the seabed, perhaps a behavioral relic of seasonal movements in an ice age.

The most important "shrimp" families of the suborder Dendrobranchiata are the penaeids and sergestids, the Penaeidae including the most commercially important shrimps in the world (genus *Penaeus*), particularly dominating seas of southern latitudes.

Among the "shrimp" families in the much larger suborder Pleocyemata, the Stenopodidae comprise cleaner shrimps, which remove ectoparasites from the bodies of fish. Carideans (infraorder Caridea) are typified by possessing a second abdominal segment of which the lateral edges overlap the segments to either side. They are the dominant "shrimps" of northern latitudes although present throughout the world's oceans, from the intertidal zone down to the deep sea. Some shrimps (eg *Macrobrachium* species) complete their entire life cycles in freshwater, but many shrimps living in rivers return to estuaries to breed and release their zoea larvae.

In addition to the totally pelagic species, many shrimps are essentially bottom dwellers, only swimming intermittently. In adult

Migrations Along the Seabed

The spiny lobsters (also occasionally called rock lobsters or marine crayfish) lack claws but have defensive spines on the carapace and antennae. Members of the family (Palinuridae) spend their days in crevices in rock or coral, emerging by night to forage for invertebrate prey. They return from their nightly wanderings to one of several dens within a feeding range of hundreds of yards, and after several weeks may move several miles to a new location. Among the best known are two American species, *Panulirus argus*, found in the shallow seas off Florida and the Caribbean, and *P. interruptus*, which lives off California. The European spiny lobster is *Palinurus elephas*, the generic name a curious anagram of that of the American species.

Many spiny lobsters take part in spectacular mass migrations. *Panulirus argus*, for example, will abandon its normal behavior in the fall and as many as 100,000 individuals move south, by both day and night, in single files of up to 60 individuals. They may cover 9.3mi (15km) a day and travel for 31mi (50km) at depths between 10 and 100ft (3–30m). The lobsters may be primed to migrate by annual changes in temperature and daylight period,

but the immediate trigger is a sharp temperature drop associated with a fall storm—usually the first strong squall of the winter. The spiny lobsters maintain alignment in the queue by touch but may respond initially by sight, recognizing the rows of white spots along the abdomens of their companions.

Mud shrimps, mud lobsters
Infraorder Thalassinidea

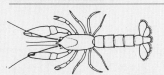

Seven families; bear a carapace up to 3.5in (9cm) long. Live in shallow water in deep burrows in sand or mud. Eg *Callianassa subterranea*.

Anomurans
Infraorder Anomura
Thirteen families.

Hermit crabs
Superfamily Paguroidea

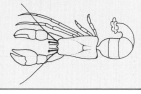

Three families; carapace to 1.6in (4cm); inhabiting marine gastropod shells; scavengers feeding on detritus. Includes *Pagurus bernhardus*, *Lithodes* species (**stone crabs**).

Land hermit crabs
Superfamily Coenobitoidea
Four families with carapace to 7.5in (19cm); on shore, land, in shells; scavenging detritivores. Includes **Coconut crab** (*Birgus latro*).

Squat lobsters, porcelain crabs
Superfamily Galatheoidea

Four families with carapace to 1.6in (4cm); marine, in holes, under stones; scavenging detritivores. Eg **Porcelain crab** (*Porcellana platycheles*).

Mole crabs
Superfamily Hippoidea

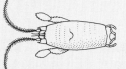

Two families with carapace to 2in (5cm), marine sand burrowers, filter feeders. Eg **Mole crab** (*Emerita talpoida*).

Crabs, spider crabs
Infraorder Brachyura

Forty-seven families with carapace up to 18in (45cm); mostly bottom dwellers, walking on seabed, but some parasitic/commensal with fish, some burrowers, some swimmers; predaceous scavengers. Includes 4 families of **spider crabs**, eg *Macrocheira kaempferi* (family Majidae); **Edible crab** (*Cancer pagurus*) and **Dungeness crab** (*C. magister*) (family Cancridae); **helmet crabs** (family Corystidae), eg *Corystes cassivelaunus*; **swimming crabs** (family Portunidae), eg **Blue crab** (*Callinectus sapidus*), **Henslow's swimming crab** (*Polybius henslowi*), and **Shore** or **Green crab** (*Carcinus maenas*); **Chinese mitten crab** (*Eriocheir sinensis*), *Sesarma* species (family Grapsidae); **pea crabs** (family Pinnotheridae), eg *Pinnotheres* species; **ghost crabs** (*Ocypode* species) and **fiddler crabs** (*Uca* species) (family Ocypodidae); and **coral gall crabs** (family Hapalocarcinidae), eg *Hapalocarcinus* species.

shrimps the thoracic pereiopods are responsible for walking (and/or feeding) and the five pairs of abdominal pleopods for swimming. Flexion of the abdomen is occasionally used for rapid escape. Shrimp zoea larvae have two-branched thoracic appendages, the exopods being used for swimming; the pleopods take over the swimming function in post-larval and adult stages as the exopods are reduced. By the adult stage the "walking" pereiopods are single-branched.

Most pelagic shrimps are active predators feeding on crustaceans of the zooplankton, such as krill and copepods. Bottom-dwelling species are usually scavengers, but range from catholic carnivores to specialist herbivores.

Shrimps usually have distinct sexes, although in some species, including *Pandalus borealis*, some of the females pass through an earlier male stage (protandrous hermaphroditism). Successful copulation usually requires molting by the female immediately before mating, when spermatophores are transferred from the male. Eggs are spawned between two and 48 hours later and are fertilized by sperm from the spermatophore. The eggs of penaeids are shed directly into the water, but in most shrimps the eggs are attached to bristles on the inner branches (endopods) of pleopods 1–4 of the female. The incubation period usually lasts between one and four months, during which time the female does not molt. Most shrimp eggs hatch as zoeae and pass through several molts over a few weeks, to post-larval and adult stages.

▲ **The squat lobster** *Galathea strigosa*, not a close relation of true lobsters, is a colorful crustacean from the lower shore and shallow water.

◄ **"House hunting,"** an important operation for this growing hermit crab seeking a larger home.

▶ **On the attack** OVERLEAF a Costa Rican ghost crab (*Ocypode* species) tackles a hatchling turtle, exposed in its efforts to put to sea for the first time.

House-hunting Hermits

Hermit crabs typically occupy the empty shells of dead sea snails, thereby gaining protection while at the same time retaining their mobility. They are able to discriminate if offered a selection of shells, and will differentiate between shells of different sizes and species, to choose the one that fits the body most closely. Hermit crabs (family Paguridae) change shells as they grow, although in some marine environments there may be a shortage of available shells, and a hermit crab may be restricted to a shell smaller than would be ideal. Some hermit crabs are aggressive and will fight fellows of their own species to effect a shell exchange; aggression is often increased if the shell is particularly inadequate.

Hermit crabs may encounter empty shells in the course of their day-to-day activity but the vacant shell is usually "spotted" by sight; the hermit crab's visual response increases with the size of an object and its contrast against the background. The hermit crab then takes hold of the shell with its walking legs and will climb onto it, monitoring its texture. Exploration may cease at any time, but if the shell is suitable the hermit crab will explore the shell's shape and texture by rolling it over between the walking legs and running its opened claws over the surface. Once the shell aperture has been located, the hermit crab will explore it by inserting its claws one at a time, occasionally also using its first walking legs. Any foreign material will be removed before the crab rises above the aperture, flexes its abdomen and enters the shell backward. The shell interior is monitored by the abdomen as the crab repeatedly enters and withdraws. The crab will then emerge, turn the shell over and re-enter finally.

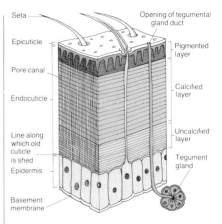

▲ ▼ **Cuticle of crustaceans** provides a tough protective layer that serves also as an external skeleton (exoskeleton). The cuticle is rich in chitin, a polysaccharide similar in structure to cellulose of plants, and is secreted by the underlying single layer of living cells, the epidermis. The endocuticle of a decapod comprises an outer pigmented, and inner calcified and uncalcified layers. The thin outermost epicuticle is secreted by tegumental glands.

The cuticle, at first soft and flexible, is progressively hardened inward from the epicuticle by the deposition of calcium salts and by sclerotization (tanning), involving the chemical cross-linking of cuticular proteins to form an impenetrable meshwork.

A rigid exoskeleton prevents a gradual increase in size, and so a crustacean grows, stepwise, in a series of molts involving the secretion of a new cuticle by the epidermis and the shedding of the old. Before a molt, food reserves are accumulated and calcium is removed from the old cuticle and much of its organic material resorbed. It is then split along lines of weakness as the crustacean swells (usually by an intake of water), before hardening of the new cuticle. The rest of the old exoskeleton may then be eaten to reduce energy losses BELOW (*Geograpsus grayii*).

During and immediately after molting the temporarily defenseless crustacean hides to avoid predators. Some adult crabs cease molting, but many adult crustaceans molt throughout life.

Most shrimps are mature by the end of their first year and typically live two to three years. (See also pp106–107.)

Lobsters and freshwater crayfish

Lobsters and freshwater crayfish belong to a group known historically as the macrurans ("large-tails") but now divided into three infraorders—the Astacidea (lobsters, freshwater crayfish, scampi), the Palinura (spiny and Spanish lobsters) and the Thalassinidea (mud lobsters and mud shrimps). All have well-developed abdomens, compared with, for example, anomurans or brachyurans. For the group's importance as a source of food for humans, see pp106–107.

Lobsters and freshwater crayfish walk along the substrate on their four back pairs of single-branched thoracic legs (pereiopods). The first pair of pereiopods is adapted as a pair of formidable claws for both offense and defense. The abdomen bears pleopods, but these cannot move the heavy body and have become variously adapted for functions which include copulation or egg-bearing.

Lobsters are marine carnivorous scavengers, usually living in holes on rocky bottoms. The commercially important American lobster reaches 2ft (60cm) in length and weighs up to 48lb (22kg). Lobsters are very long lived and they may survive 100 years.

Freshwater crayfish, however, are more omnivorous. There are more than 500 species of freshwater crayfish, mostly about 4in (10cm) long, and since they typically

require calcium they are often restricted to calcareous waters. Because of their marine ancestry, their internal fluids have an osmotic pressure higher than that of the surrounding medium and water will enter by osmosis across gill and gut membranes. The cuticle covering the rest of the body is made impermeable by tanning and calcification. The convoluted tubule of each antennal (green) gland is very long and resorbs ions from the primary urine (filtered from the blood at the end sac of the gland). A dilute urine, one-tenth the osmotic pressure of the blood, may be produced, eliminating water that has entered osmotically. Any salts still lost with the urine are replaced by ions actively taken up by the gills.

Reproduction involves pairing, in which either sperm from the male flows down along the grooves on the first pleopods into a seminal receptacle in the female, or a spermatophore is transferred. Fertilized eggs are incubated on the pleopods of the female and hatch as mysis larvae bypassing in the egg the stages equivalent to nauplius and zoea.

Squat lobsters and hermit crabs

Squat lobsters, hermit crabs and mole crabs are part of a collection of probably unrelated groups intermediate in structure and habit between lobsters and crabs. The abdomen is variable in structure—often being asymmetrical or reduced and held in a flexed position (anom-ura: odd-tailed—hence the scientific name Anomura given to these crustaceans). The fifth pair of pereiopods is turned up or reduced in size.

Hermit crabs have probably evolved from ancestors which regularly used crevices for protection, eventually specializing in the use of the discarded spiral shells of gastropod mollusks. The abdomen is adapted to occupy the typically right-handed spiral of such shells although rarer left-handed shells may also be used. The pleopods are lost at least from the short side of the asymmetric abdomen, but those on the long side of females are adapted to carry fertilized eggs. At the tip of the abdomen, the uropods are modified to grip the interior of the shell posteriorly, and toward the front of the animal thoracic legs may be used for purchase. One or both claws can block up the shell opening. The shell offers excellent protection and still affords mobility (see box, p100).

Not all species of hermit crabs live in gastropod shells. Some occupy tusk shells, coral, or holes in wood or stone. The hermit crab *Pagurus prideauxi* lives in association with a sea anemone, *Adamsia carciniopados*,

whose horny base surrounds the crab's abdomen and greatly overflows the originally occupied shell. This hermit crab avoids the risk of being eaten when transferring from one shell to another, for the protecting anemone moves simultaneously with its associate. The crab is protected by the stinging tentacles of the anemone, which in turn profits from food particles which its crustacean partner releases into the water during feeding.

Hermit crabs live as carnivorous scavengers on sea bottoms ranging from the deep sea to the seashore. They have also taken up an essentially terrestrial existence in the tropics. Land hermit crabs range inland from the upper shore, often occupying the shells of land snails. The Coconut crab has abandoned the typical hermit crab form, and appears somewhat crab-like, but with a flexed abdomen. It lives in burrows and holes in trees, which it can climb, feeding on carrion and vegetation, and can drink. Land hermit crabs have reduced the number of gills that would tend to dry out in the air and collapse under surface tension. The walls of the branchial chamber are richly supplied with blood, which enables them to act as lungs, and some species have accessory respiratory areas on the abdomen enclosed in the humid microclimate provided by the shell. Land hermit crabs are not fully terrestrial, for they have planktonic zoea larvae. The adults must return therefore to the sea for reproduction.

Squat lobsters are well named, for they have relatively large symmetrical abdomens flexed beneath the body. They typically retreat into crevices. Porcelain crabs are anomuran relatives of squat lobsters which look remarkably like true crabs. Mole crabs are anomurans that by flexing the abdomen burrow into the sand when waves break low on warm sandy shores. The first antennae form a siphon channeling a ventilation current to the gills and the setose (bristly) second antennae filter plankton.

True crabs

There are 4,500 species of true crabs which all possess a very reduced symmetrical abdomen held permanently flexed beneath the combined cephalothorax (brachyuran: short-tailed). The terminal uropods are lost in both sexes; four pairs of pleopods are retained on the female's abdomen to brood eggs and in the male only the front two pairs remain to act as copulatory organs. The crabs have a massive carapace extended at the sides, and the first of the five pairs of thoracic walking legs is adapted as large claws. Typically crabs are carnivorous walkers on the sea bottom.

The reduction of the abdomen has brought the center of gravity of the body directly over the walking legs, making locomotion very efficient and potentially rapid. The sideways gait assists to this end. The shape of a crab is therefore the ultimate shape in efficient crustacean walking. Mainly as a result, the brachyurans have enjoyed an explosive adaptive radiation since their origin in the Jurassic era (195–135 million years ago) and various anomurans (for example mole crabs and porcelain crabs) approach or duplicate the crab-like form.

Crabs live in the deep sea and extend up to and beyond the top of the shore. The blind crab *Bythograea thermydron* is a predator in the unique faunal communities surrounding deep-sea hydrothermal vents, regions of activity in the earth's crust which emit hot sulfurous material 1.6mi (2.5km) below the surface of the sea. On the other hand, crabs of the family Ocypodidae, such as the burrowing ghost and fiddler crabs, live at the top of tropical sandy and muddy shores and the distribution of the genus *Sesarma* extends well inland. Crabs have also invaded the rivers—the common British Shore crab extends up estuaries—and in the tropics crabs of the family Potamidae and some Grapsidae are truly freshwater in habit.

Most crabs burrow to escape predators—descending back-first into the sediment, and several typically remain burrowed for long periods. Of these the Helmet crab has long second antennae which interconnect with bristles to form a tube for the passage of a ventilation current down into the gill chamber of the buried crab. Some crabs have become specialist swimmers, with the last pair of thoracic legs adapted as paddles. The more terrestrial crabs like the ghost crabs are very rapid runners—their speed and nighttime activity contributing to their common name. Other crabs, particularly spider crabs (see box), cover themselves with small plants and stationary animals (eg sponges, anemones, sea mats) for protective camouflage. Pea crabs may live in the mantle cavities of bivalve mollusks, feeding on food collected by the gills of the host, and female coral gall crabs become imprisoned by surrounding coral growth. The female crab is left with just a hole to allow entry of plankton for food and the tiny male for reproduction.

Most crabs are carnivorous scavengers although the more terrestrial ones may eat

▲ **Quest in the trees:** the Blue land crab *Cardisoma guanhumi* can climb trees in search of prey.

▶ **Getting to grips:** the large nippers of the fiddler crab are used for signaling and ritual combat. Here two males of the genus *Annulipes* test their strength.

▼ **Buried alive.** The ghost crab *Ocypode ceratophthalma* seeks refuge from predators by burrowing in the sand. Only its large eyes protrude to keep watch.

Japanese Spider Crabs

Spider crabs belong to a family of true (brachyuran) crabs with long legs and a superficial resemblance to spiders. Included in their number is the world's largest crustacean, the giant spider crab (*Macrocheira kaempferi*), whose Japanese name is Takaashigani—the tall-leg crab.

This remarkable animal may measure up to 26.5ft (8m) between the tips of its legs when these are splayed out on either side of the body. The claws of such a beast may be 10ft (3m) apart when held in an offensive posture. The main body (cephalothorax) of the crab is, however, relatively small and would not usually exceed 18in (45cm) in length or 12in (30cm) in breadth.

Giant spider crabs have a restricted distribution off the southeast coast of Japan. They live on sandy or muddy bottoms between 100–165ft (30–50m) deep. They have poor balance and so live in still waters, hunting slow-moving prey that includes other

crustaceans as well as echinoderms, worms and mollusks. Little detail is known of their life history, but they probably pass through zoea and megalopa larval stages before they metamorphose into juvenile crabs. Large specimens are probably more than 20 years old.

The crabs are caught relatively infrequently, but they are used for food. Because of their size they command respect, and their nippers can inflict a nasty wound.

The existence of the giant crabs was first reported in Europe by Engelbert Kaempfer, a physician for the Dutch East India Company who visited Japan in 1690, in his *History of Japan* published in English in 1727 (although he himself died in 1716). The crabs were given their Latin name in honor of Kaempfer in the 19th century by C. J. Temminck, the director of the Leiden Museum, which received much of the natural history collections of the Dutch East India Company.

plant matter. The fiddler crabs process sand or mud in their mouthparts, scraping off the nutritious microorganisms with specialized spoon- or bristle-shaped setae.

Reproduction involves copulation. The second pair of pleopods on the male's abdomen acts like pistons within the first pair to transfer sperm to the female for storage. The eggs are fertilized as they are laid and are held under the broad abdomen of the female. They hatch as pre-zoea larvae, molting immediately to the first of several planktonic zoea stages. Zoea larvae have a full complement of head appendages, but only the first two pairs of thoracic appendages (destined to become the first two or three pairs of adult maxillipeds), used by the zoea for swimming. After several molts the megalopa stage is reached, with abdominal as well as thoracic appendages. This settles on the sea bottom and metamorphoses into a crab. PSR

Economic Importance of Crustaceans
Human food and foulers of vessels

The economic importance of crustaceans is twofold—their commercial value as food items and the costs caused by their effects as foulers of ships or coastal structures.

Crustaceans, especially the decapods (shrimps, lobsters, crabs), are important as food products, whether cropped from the wild or reared by aquacultural processes. Whale krill of the Antarctic Ocean are now being harvested and processed as food, not only for man, but more particularly for agricultural livestock. In the Gulf of Mexico and off the southeastern USA, fishing boats trawl for the Brown shrimp which, with the White shrimp and the Pink shrimp, make up the world's largest shrimp fishery. In Southeast Asia, in Indonesia and the Philippines and in Taiwan, shrimps of the same genus *Penaeus* have been reared for food in brackish ponds for 500 years. Popular species are the Banana prawn, the Indian prawn, the Giant tiger prawn, the Green tiger prawn and the Yellow prawn. The shrimps are trapped in pools at high tide and cultured for several months, often with mullet or milkfish. In Singapore ponds are usually constructed in mangrove swamps and in India rice paddies are used. Typical yields vary from 1,650 to 8,700lb of edible shrimp per acre (300–1,600kg/hectare). In Japan there is a recently developed intensive culture of the Kuruma shrimp. Egg-bearing females are supplied by fishermen and shrimps are reared over two weeks, from eggs through successive larval stages with differing food requirements, to postlarvae before transfer to production ponds. Harvesting takes place 6–9 months later at yields which may attain 32,600 lb per acre (6,000kg/hectare).

In British waters the Caridean shrimp (also called the Pink shrimp) is fished by beam trawl in spring and summer in the Thames estuary, the Wash, Solway Firth and Morecambe Bay—the Thames fishery dating back to the 13th century. The Brown shrimp is fished off Britain, Germany, Holland and Belgium; boats work a pair of beam trawls and total landings reach 30,000 tons a year. *Palaemon adspersus* is taken off Denmark and off south and southwest Britain.

Another caridean, the tropical Indo-Pacific freshwater prawn *Macrobrachium rosenbergi*, is a giant reaching 10in (25cm) long; it is therefore attractive to aquaculturists, but its larvae are difficult to rear and dense populations do not occur naturally.

Lobsters and crabs are usually caught in traps, enticed by bait to enter via a tunnel of decreasing diameter protruding into a wider chamber. Most are predators or

scavengers: lobsters prefer stinking bait but crabs are attracted to fresh fish pieces. Crab or lobster meat usually consists of muscle (white meat) extracted from the claws and legs, and lobster abdominal muscle is also used. The hepatopancreas (brown meat) may be taken in some species. The meat is processed as a meat paste or canned, or the crustacean may be sold fresh or frozen.

The Dublin Bay prawn or Norway lobster supplies scampi—strictly the abdominal muscle. It burrows in bottom mud and is collected by trawling. Most lobsters, however, are caught in lobster pots. American lobsters and European lobsters are of commercial value in temperate seas but are replaced by spiny lobsters in warmer waters: *Panulirus argus* and *P. interruptus* are trapped off Florida and California respectively, *P. versicolor* throughout the Indo-Pacific and *Jasus* species off Australia.

▲ **Four different species** of *Pandalus* make up the principal elements of the shrimp catch from southwest Alaska.

◄ **Harvest of the sea:** commercial shrimp boat unloading at Kachemak Bay, southwest Alaska.

Freshwater crayfish are consumed with enthusiasm in France. The habit has transferred to Louisiana, USA, where between 880,000 and 1,960,000lb (from 400,000 to 800,000kg) of wild Red and White crayfish are trapped each year, and a further 2.6 million lb (1.2 million kg) reared in artificial impoundments.

The Edible crab (*Cancer pagurus*), is taken in pots off European coasts. The British annual catch reaches 6,500 tons, boats laying out 200–500 pots daily in strings of 20–70 buoyed at each end. The crabs are usually sold to processing factories, to be killed by immersion in freshwater or by spiking, boiled in brackish water and then cooled to room temperature to set the meat, which is extracted by hand picking or with compressed air. The related Dungeness crab is fished off the west coast of North America, and the Blue crab is taken off the east coast by trap or line or by fishing with a trawl net.

On the negative side, one year's growth of fouling organisms can increase the fuel costs of ships by 40 percent, and barnacles are the most important foulers. Their presence impedes the smooth flow of water over a ship's hull and their shell plates disrupt paint films, so enhancing corrosion. Barnacles are cemented to the substrate and necessitate expensive dry docking for removal. Barnacles have swimming larvae that disperse widely, ready to alight on passing ships. Barnacles and tube-building crustaceans may also clog water-cooling intake pipes of industrial installations by the coast.

The commercial importance of wood-boring crustaceans, such as the gribbles, has decreased since concrete has replaced wood as a major pier construction material.

HORSESHOE CRABS AND SEA SPIDERS

Phylum: Chelicerata (part)

Horseshoe or king crabs
Class: Merostomata.
Order: Xiphosura.
Four species in 3 genera.
Distribution: marine, Atlantic coast of
N America (genus *Limulus*), coasts of SE Asia
(*Tachypleus, Carcinoscorpius*).
Fossil record: first appear in the Devonian
period, 400–350 million years ago.
Size: larvae about 0.4in (1cm) long, adults up
to 2ft (60cm) long.

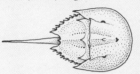

Features: pair of pincer-like mouthparts
(chelicerae) in front of mouth; carapace covers
the back of the prosoma (front portion of body)
and part of the opisthosoma (rear portion of
body); prosoma has 6 pairs of appendages:
chelicerae and 5 pairs of walking legs;
opisthosoma bears 6 pairs of flattened limbs (a
modified genital operculum or lid, and 5 pairs
of leaf-like or lamellate gillbooks for respiration)
and a rear spine; excretion is by coxal glands
at limb bases; circulatory system well
developed; sexes distinct, with external
fertilization; larva has three divisions, like a
trilobite.

Sea spiders
Class: Pycnogonida.
Order: Pantopoda.
Five hundred species in 70 genera and 8
families.
Distribution: marine, worldwide, from shores to
ocean depths.
Fossil record: appear first in the Devonian
period 400–350 million years ago.
Size: narrow, short body 0.04–2.5in (0.1–6cm)
long.

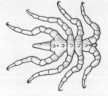

Features: cephalon has a tubular proboscis
with mouth at tip, paired chelicerae and palps;
usually 4 pairs of long legs on protuberances at
the side; adult heart tubular, within blood-
containing body cavity (hemocoel); there are
no excretory, respiratory or osmoregulatory
organs; sexes distinct, with external
fertilization; extra, egg-carrying, legs well
developed on males; males brood eggs which
hatch into protonymphon larvae.

Families: Ammotheidae (eg *Achelia* species);
Colessendeidae; Endeidae; Nymphonidae (deep-
sea, including 100 species of *Nymphon*);
Pallenidae; Phoxichilidiidae; Pycnogonidae;
(eg *Pycnogonum* species); Tanystylidae.

MOST chelicerates are land creatures, such as spiders, scorpions, harvestmen and the parasitic mites and ticks. The aquatic chelicerates, however, are by no means so well known. They consist of two apparently dissimilar groups—the horse-shoe or king crabs, "living fossils," and the sea spiders, "all legs and no body."

All chelicerates are typified by having a pair of pincer-like mouthparts, known as chelicerae, in front of the mouth opening, where insects and crustaceans have one and two pairs of antennae respectively. In addition chelicerates have no biting jaws and in most (including the horseshoe crabs but not sea spiders) two distinct parts of the body are recognizable, the prosoma (front part) and opisthosoma (rear part). Since they fit this description and in spite of their common names and sea-dwelling habit, horseshoe crabs are therefore more closely related to terrestrial arachnids than to crustaceans. They have remained more or less unchanged for over 300 million years. The affiliations of the sea spiders are less clear, their idiosyncratic body shape probably having been much altered during evolution. It is probable that sea spiders should be included in the chelicerates, and some authorities believe them actually to be arachnids, and close to their terrestrial counterparts.

Horseshoe crabs or king crabs have a pro-tective hinged carapace whose domed shield of horseshoe form covers the prosoma and the first part of the opisthosoma; a long caudal spine protrudes behind. They have compound eyes on the carapace and median simple eyes. Beneath the dark brown carapace lie the chelicerae and five pairs of walking legs, comparable in evolution to the chelicerae, pedipalps and four pairs of walk-ing legs of spiders.

Horseshoe crabs live on sandy or muddy bottoms in the sea, plowing their way through the upper surface of the sediment. During burrowing, the caudal spine levers the body down while the fifth pair of walking legs acts as shovels, the form of the carapace facilitating passage through the sand. The animal also uses the caudal spine to right itself if accidentally turned over.

Horseshoe crabs are essentially scaveng-ing carnivores. There are jaw-like exten-sions on the bases of the walking legs, used to trap and macerate prey, such as clams and worms, before it is passed forward to the mouth. The stout bases of the sixth pair of legs can also act like nutcrackers to open the shells of bivalves which are seized during burrowing.

The appendages on the rear part (opis-thosoma) are much modified. The first pair forms a protective cover over the remainder, each of which is expanded into about 150 delicate gill lamellae resembling the leaves of a book, in an adaptation unique to horse-shoe crabs. The movement of the appenda-ges maintains a current over the respiratory surfaces of the gill books which are well sup-plied with blood, which in turn drains back to the heart for pumping around the body. Small horseshoe crabs can swim along upside down, using the gill books as paddles.

At night at particular seasons of the year, male and female horseshoe crabs congre-gate at the intertidal zone for reproduction. The female lays 200–300 eggs in a depres-sion in the sand, to be fertilized by an attendant clasping male. The eggs hatch after several months as trilobite larvae, so called because of their similarity to the trilobites of the fossil record of the Paleozoic era (600–225 million years ago). Initially 0.4in (1cm) long, the larvae develop to reach maturity in their third year.

Sea spiders are exclusively marine, and are to be found from the intertidal zone to the deep sea. They have an exaggerated hang-ing stance, like larger examples of their ter-restrial namesakes, and are typically to be found straddling their stationary prey, which includes hydroid polyps, sponges and sea mats. Sea spiders grip the substrate with claws as they sway from one individual of their colonial prey to another without shift-ing leg positions. Although they mostly move slowly over the seabed, many sea spiders are capable swimmers.

Sea spiders typically have four pairs of long legs arising from lateral protuberances along the sides of the small, narrow trunk. There are some ten- and twelve-legged species. The paired chelicerae and palps border the proboscis. Sea spiders usually feed either by sucking up their prey's body tissues through the proboscis or by cutting off pieces of tissue from the prey with the chelicerae, then transferring them to the mouth at the tip of the proboscis. A few sea spiders feed on detritus.

In addition to the walking legs, there is a further pair of small legs, which is particularly well developed in males. As eggs are laid by the female they are fertilized and collected by the male, which cements them on to the fourth joint of each of its small egg-bearing (ovigerous) legs, where they are brooded. The eggs hatch later as protonym-phon larvae and develop through a series of molts, adding appendages to the original three larval pairs, the forerunners of the chelicerae, palps and ovigerous legs.

Body colors are variable, generally white/transparent, but red in deep-sea species. The high surface-area-to-volume ratio of the narrow body means that it is only a short distance to the outside from anywhere in the body and respiratory gases and other dissolved substances can be moved efficiently by diffusion. There is therefore no necessity for specialized excretory, osmoregulatory or respiratory organs. The lack of storage space in the body does, however, mean that reproductive organs and gut diverticula have to be partly accommodated in the relatively large legs.   PSR

◄ **Nursing father:** the male sea spider *Nymphon gracile* has special ovigerous legs to carry the fertile eggs.

▼ **Not from outer space** but from distant time, "living fossil" horseshoe crabs (*Limulus* species) mating on the shore.

WATER BEARS AND TONGUE WORMS

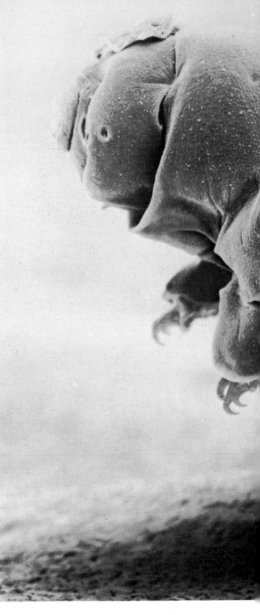

Phyla: Tardigrada and Pentastomida

Water bears

Phylum: Tardigrada.
Four hundred species in 3 classes and 15 families.
Distribution: worldwide, particularly on damp moss, also in soil, freshwater and marine sediments.
Fossil record: species from the Cretaceous period, 135–65 million years ago.
Size: microscopic to minute—under 0.05in (50μm to 1.2mm)

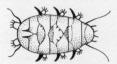

Features: fat-bodied, with a soft, chitinous cuticle, periodically molted; 4 pairs of stout, unjointed, clawed legs; sucking mouthparts with paired stylets; simple gut; segmented muscles; nerve cord on underside, with swellings (ganglia); no circulatory or respiratory organs; 3 "Malpighian" glands are possibly excretory; probably a fixed number of cells; only sense organs are 2 eyespots; remarkable suspended animation (anabiosis) during desiccation; sexes distinct, internal fertilization (some asexual reproduction).

Comprises classes Eutardigrada (2 orders, 6 families), including Macrobiotidae (eg *Macrobiotus* species); Heterotardigrada (2 orders, 8 families), including Batillipedidae (eg *Batillipes* species) and Echiniscidae (eg *Echiniscus*); and Mesotardigrada (1 order), including family Thermozodiidae.

Tongue worms

Phylum: Pentastomida.
Ninety species in 6 families.
Distribution: worldwide, parasitizing vertebrates.
Size: 0.1–5.5in (3mm–14cm).

Features: worm-like parasites in respiratory tracts of carnivorous vertebrates, particularly reptiles; larvae usually in intermediate herbivorous vertebrate host; chitinous cuticle; 5 anterior protuberances, the first bearing sucking mouth, the others hooks; 2 pairs of rudimentary limbs in larvae, and in adults of genus *Cephalobaena*; straight gut; reduced segmented nervous system; no circulatory, respiratory or excretory organs; sexes distinct, internal fertilization.

Families: Linguatulidae (**tongue worms**), including fluke-like *Linguatula* species; Porocephalidae, including long, ringed *Porocephalus* species, *Kiricephalus* species; Raillietiellidae, including *Cephalobaena* species; Reighardiidae; Sambonidae; Sebekidae.

WATER BEARS and tongue worms are relatively obscure animals which have similarities to arthropods, particularly in having a chitinous cuticle. Some zoological authorities go as far as to include each group in one of the three main arthropod phyla, water bears in the uniramians and tongue worms in the crustaceans. Here water bears and tongue worms are assigned to their own phyla, the Tardigrada and the Pentastomida.

The **water bears** or **tardigrades** (literally "slow steppers") are minute fat-bodied animals said to resemble bears, and often referred to as water bears, moss bears or bear animalcules. They are to be found in the water film on damp moss and also occur in soil, freshwater and marine sediments.

Water bears move by slow crawling, attaching with claws at the end of each leg. In addition to the four pairs of short unjointed legs, there are further signs of segmentation in the muscles and the nervous system. In *Echiniscus* species the cuticle is arranged in segmental plates. Most water bears feed by piercing plant cells with two sharp stylets and sucking out the contents via a muscular pharynx. Some feed on detritus and others are predatory. There are two, probably salivary, glands leading into the mouth. Defecation may be associated with molting, as in *Echiniscus* which leaves its feces in its cast cuticle.

The three "Malpighian" glands leading into the gut are believed to serve an excretory function. There are no respiratory organs, blood vessels, nor heart, because diffusion is sufficient for transport of essential foods in such small animals. The nerve cord along the underside is well organized with ganglia, and the sense organs include two eyespots and sensory bristles.

Water bears have separate sexes, the females usually being the more numerous. However, males have yet to be discovered in certain species, and in other species the females breed asexually (parthenogenesis). In general, reproduction involves copulation with subsequent internal fertilization; some females store transferred sperm in seminal receptacles. Between one and 30 eggs are laid at a time. These may be thin-shelled and hatch soon after laying, or thick-shelled to resist hazardous environmental conditions. Newly hatched young resemble adults and grow by increasing the size, but not the apparently fixed number of their constituent cells—a feature perhaps associated with their very small size. Legs increase from two pairs to four before adulthood. Maturity is reached in about 14 days and they live between three and 30 months, passing through up to 12 molts.

Water bears are remarkable in withstanding extreme conditions. They can survive desiccation and, experimentally at the other extreme, immersion into chemicals or liquid helium at −337°F. The tardigrade enters a state of dormancy that may last years, reviving when conditions improve.

The phyletic position of the water bears is controversial, for they resemble other groups as well as arthropods. The **tongue worms** or **pentastomids**, however, are clearly arthropodan—disagreement existing over whether they should be placed in their own phylum or included in the crustaceans. Tongue worms are worm-like parasites in the respiratory tracts of carnivorous vertebrates, typically reptiles, although six species are found in mammals including man and two in birds. The life cycle usually involves a herbivorous vertebrate intermediate host for the developing larvae, such as a fish or a rabbit.

▲ **Blown-up water bear.** This highly enlarged photograph shows clearly the external morphology of the water bear *Macrobiotus richtersi* (× 700).

▼ **Digesting its last meal.** The plant food in the gut of this tardigrade can be clearly seen.

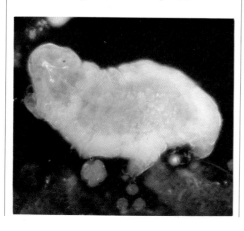

The front end of a tongue worm has five short protuberances bearing, respectively, the mouth and four hooks used by the animal to anchor itself and in feeding. (Once it was thought that there was a mouth on each of the five—hence the name pentastomid.) The mouth is modified for sucking the blood of the host. Some species have superficial ring-like markings along the body, enhancing the worm-like appearance. The worm shape is an adaptation to life in confined passages which must not be blocked; otherwise the host will be suffocated.

The body is covered in a chitinous cuticle which is molted during larval development. Most larvae have two pairs of appendages, and these are retained in adults in the primitive genus *Cephalobaena*. The adults move little, and the nervous system is reduced, as befits a parasite. There is no heart, and respiratory and excretory organs are also absent.

There are distinct sexes, and fertilization is internal. Fertilized eggs are released in the host's feces or by the host sneezing, and lie in vegetation before being ingested by a herbivore. In the case of *Linguatula serrata* the eggs are eaten by a rabbit or hare and the larvae hatch out under the action of digestive juices. The larva bores through the gut wall and is carried in the blood to the liver. Here it forms a cyst and grows through a series of molts to approach adult form. If the rabbit is eaten by a dog or wolf, the larva is released and passes up the dog's esophagus to the pharynx, and so to its adult location in the host's nasal cavity.

Porocephalus crotali lives in snakes with mice as intermediate hosts, and other tongue worms live in crocodiles with fish intermediate hosts. It has been suggested that pentastomids may be modified crustaceans originally parasitic on fish in the manner of the modern genus *Argulus* of crustacean fish lice. It is a plausible step for the parasite to have been stimulated to leave this host when eaten by a crocodile, to take up residence in the mouth of the predator.　　PSR

Mollusks
Phylum: Mollusca

Spiny-skinned animals
Phylum: Echinodermata

Chordates
Phylum: Chordata

Sea squirts
Subphylum: Urochordata (or Tunicata)

Lancelets
Subphylum: Cephalochordata (or Acrania)

Vertebrates (not in this volume)
Suphylum: Craniata (or Vertebrata)

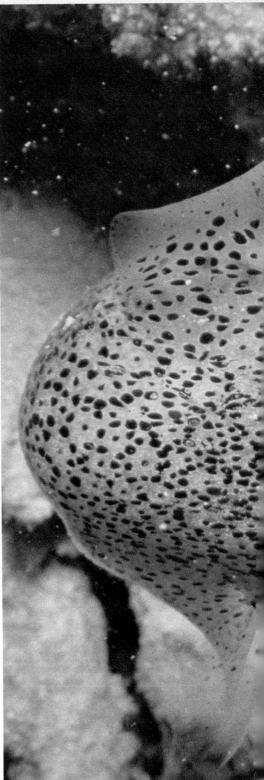

W E think of most animals as having distinct front and rear ends, a definite head, and left and right sides. Not all animal phyla fit this pattern, and some have a body plan and architecture which zoologists still do not fully understand. Among these phyla are the mollusks, echinoderms and urochordates.

In the largest class of mollusks, the gastropods, the advantages of coiling and twisting (torsion) of the body in the snails and slugs are not understood. If torsion confers advantages to the larvae, as has been suggested, why does it persist in many adults, and why have some mollusks abandoned coiling and torsion in the course of evolution? Quite different in appearance, the sophisticated squids and octopuses have produced a further variation on the mol-luskan theme; in the cephalopod ("head-foot") body plan the fleshy foot has migrated to surround the head in the form of tentacles.

The spiny-skinned echinoderms, unlike the mollusks, have no head and many lack even a recognizable front end, being able to move in all directions with similar ease. Their mostly five-sided bodies have no right and left sides. Again the advantages conferred by this form have not really been satisfactorily explained. The fact that echinoderm larvae, unlike the adults, have a distinct front end, as well as left and right sides, makes matters more confusing.

The urochordates, which include the sea squirts and lancelets, are a subphylum of the great phylum Chordata. Their swimming larvae resemble primitive tadpoles, and

▲ **Sophisticate among invertebrates,** this Little cuttle displays the large head and eyes, suckered arms and the pigment spots (chromatophores) of cephalopod mollusks.

◄ **Not all starfishes have five arms,** as this *Solaster endeca* shows. Generally the five-pointed symmetry (pentamery) is superimposed on a radial body plan, as in the brittle stars that surround their relative.

have features which were crucial in identifying the chordate affinities of the group—a distinct head and tail and a rudimentary stiffening structure in the body, the notochord. However, the adults look nothing like chordates at all. The humble-looking sea squirts have a weird shape with no head.

Perhaps all these forms have at some time in their evolution served their owners to advantage, in either their larval or adult lives. Alternatively, some may not be as highly evolved as has been supposed, and the groups in question may have managed to get by in spite, rather than because, of their strange shape. In any event, today the mollusks are a major group of marine animals, significant also in freshwater and on land, where they may carry human disease or be crop pests. The echinoderms are restricted to the sea; a few are economically important as pests in shellfisheries and coral areas or as food for commercial fishes. The urochordates are a group of great interest, meriting further research that may tell us much about animal embryology and development.

AC

MOLLUSKS

Phylum: Mollusca

About 80,000–100,000 species in 7 classes.
Distribution: worldwide, primarily aquatic, in
seas mainly, some in fresh and brackish water,
some on land.
Fossil record: appear in Cambrian rocks about
530 million years ago, modern classes distinct
by about 500 million years ago; abundant
extinct nautiloids, ammonites, belemites.
Size: often smaller than eg crustaceans, but
ranging from tiny 0.04in (1mm) gastropods to
Giant squid up to 60ft (20m) long overall.

Features: body soft, typically divided into head
(lost in bivalves), muscular foot and visceral
hump containing body organs; protected by
hard calcareous shell in most species; no paired
jointed appendages; a fold of skin (mantle)
forms a cavity that protects soft parts, and may
act in defense and locomotion; mouthparts
include usually a toothed tongue (radula);
breathing usually by gills which may serve also
in filter feeding; heart usually present;
cephalopods have veins; nervous system of
paired ganglia—brain and eye of cephalopods
most highly developed of all invertebrates;
sexes typically separate (not land species),
fertilization external or by copulation; young
develop via larval stages including often a free-
floating trochophore and/or veliger before
settling to an often bottom-dwelling life.

THE diversity of mollusks encompasses food, pests, dyes, pathogens, parasites and pearls. This variety is reflected in the range of body forms and ways of life. Mollusks include coat-of-mail shells or chitons, marine, land and freshwater snails, shellless sea slugs and terrestrial slugs, tusk shells, clams or mussels, octopuses, squids, cuttlefishes and nautiluses. In addition to the 80,000–100,000 living species, the many extinct species include the ammonites and belemites. Today's mollusks occur throughout the world, living in the sea, fresh and brackish water, and on land. Apart from those that float, swim weakly (eg sea butterflies), or powerfully (eg squids), or burrow (eg clams), mollusks live either attached to or creeping over the substrate, whether seabed or ground, or vegetation.

The molluskan body is soft and is typically divided into head (lost in the bivalves), muscular foot and visceral hump containing the body organs. There are no paired jointed appendages or legs, a feature distinguishing mollusks from the arthropods. Most species have their soft parts protected by a hard calcareous shell.

Two notable features of the molluskan body are the mantle, an intucking of skin tissue that produces a protective pocket, and the toothed tongue or radula.

Mollusks have a gut with both mouth and anus, associated feeding apparatus, a blood system (generally with a heart), nervous system with ganglia, reproductive system (which in some is very complex), and excretory system with kidneys. The epidermal (skin) tissues of mollusks are generally moist and thin, and liable to drying out. Gills are present in most aquatic species which are used to extract oxygen from water. In the majority of bivalves and some gastropods, however, the gills are also used in feeding, when they strain out organisms and detritus from the water or bottom mud with minute flickering cilia on the gills. These particles are then conveyed by tracts of cilia to the mouth.

In land and some freshwater snails, the walls of the mantle cavity act as lungs, exchanging respiratory gases between the air and body. Many mollusks have a freefloating (pelagic) larval stage, but this is absent in land and some freshwater examples.

Lack of an internal skeleton has kept most mollusks to a relatively small size. Cephalopods have achieved the greatest size

▼▶ Some representative species of mollusks.
(1) A species of *Dentalium* or elephant's tusk
shell; (2in, 5cm). (2) *Solen marginatus*, a razor
shell; 4.9in (12.5cm). (3) A chiton (class
Polyplacophora). Chitons are flattened
sedentary mollusks with eight overlapping
protective shell plates. (4) *Nucella lapillus*, a
dog whelk; rocky shores of NW Europe (1.2in,
3cm high). (5) *Buccinum undatum*, the large
European whelk. It lives on sand and mud
down to about 330ft (100m) (3in, 8cm).
(6) A species of *Neopilina*, a genus of limpet-like
mollusks (1.6in, 4cm). (7) A species of *Aplysia*
or sea hare (5.9in, 15cm). (8) *Tridacna gigas*,
the largest living mollusk or giant clam (4.4ft,
1.35m). (9) A common octopus (genus
Octopus): the zenith of molluskan organization.
(10) *Falcidens gutterosus*, a chaetoderm. It lives
in mud at 130ft (40m) and below;
Mediterranean (0.6in, 1.5cm).

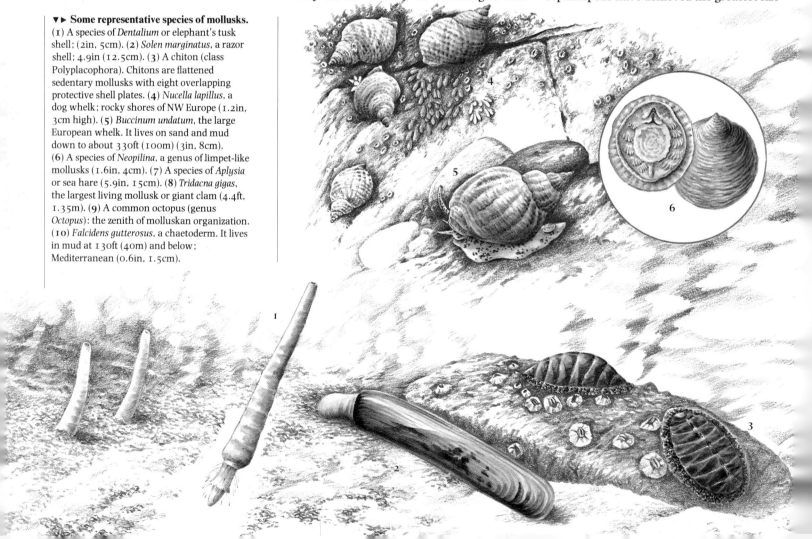

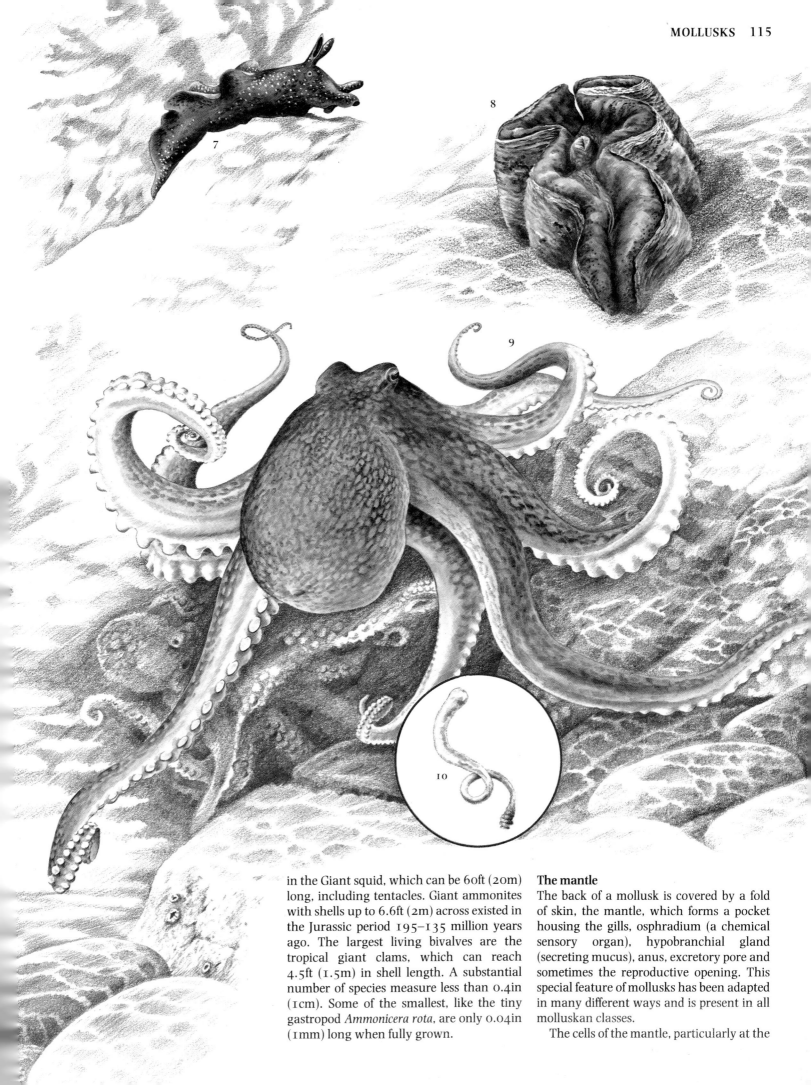

in the Giant squid, which can be 60ft (20m) long, including tentacles. Giant ammonites with shells up to 6.6ft (2m) across existed in the Jurassic period 195–135 million years ago. The largest living bivalves are the tropical giant clams, which can reach 4.5ft (1.5m) in shell length. A substantial number of species measure less than 0.4in (1cm). Some of the smallest, like the tiny gastropod *Ammonicera rota*, are only 0.04in (1mm) long when fully grown.

The mantle
The back of a mollusk is covered by a fold of skin, the mantle, which forms a pocket housing the gills, osphradium (a chemical sensory organ), hypobranchial gland (secreting mucus), anus, excretory pore and sometimes the reproductive opening. This special feature of mollusks has been adapted in many different ways and is present in all molluskan classes.

The cells of the mantle, particularly at the

thickened edge of the mantle skirt, secrete the shell and may also produce slime, acids and ink for defense. Mucus for protection and for cohesion of food particles is secreted by the gill and the hypobranchial glands. Products of the mantle can be defensive, acting to deter predators. The purple gland in the mantle of the sea hare expels a purple secretion when the animal is disturbed. Several species of dorid sea slugs or sea lemons can expel acid from glands in the mantle, while on land the Garlic snail gives off a strong aroma of garlic from cells near the breathing pore.

The mantle wall may be visible and in some sea slugs it is brightly colored and patterned—acting as either warning coloration or camouflage. The glossy and colored shells of cowries are usually hidden by a pair of flaps from the mantle.

Protection of the delicate internal organs was probably an early function of the mantle, which also provides a space into which the head and foot can be retracted when the animal withdraws into its shell. Within the mantle cavity, the gills are protected from mechanical damage from rocks, coral etc as well as from silting. At the same time the gills must have ready access to sea water from which to extract oxygen. In some mollusks, special strips of mantle tissue are developed as tubular siphons which help to separate two currents of water that pass over the gills—the inhalant and exhalant water currents. Fleshy lobes to the mantle, as in freshwater bladder snails of the genus *Physa*, may function as extra respiratory surfaces.

Fertilization of eggs may take place within the mantle cavity of bivalves, and eggs are brooded there in, for example, the small pea shells of freshwater habitats.

The versatile molluskan mantle can become muscular and serve in locomotion. Some sea slugs employ their leaf-like mantle lobes in swimming. Some scallops, such as the Queen scallop, swim by expelling water from the mantle cavity.

The radula

The radula, a toothed tongue, is typically present in all classes of mollusk except the bivalves. It is secreted continuously in a radula sac and is composed of chitin, the polysaccharide also found in arthropod exoskeletons. The oldest teeth are toward the tip: when a row of teeth becomes worn, they detach and are often passed out with the feces. A new row of teeth then moves into position. Inside the mouth is an organ, the buccal mass, which contains and operates the radula during feeding. The radula is carried on a rod of muscle and cartilage (odontophore) that projects into the mouth cavity, while further complexes of muscles and cartilage in the walls of the buccal mass operate the radula, usually in a circular motion.

The form of the radula depends on feeding habits and is used in identifying and classifying individual species. Herbivores, such as land snails and slugs, have a broad radula with many small teeth, while carnivores, including whelks, have a narrow radula with a few teeth bearing long pointed cusps. Limpets, which browse algae off rocks, have an especially hard rasping radula and a few very strong teeth in each row; they leave

▲ **Tiny black eyes** and sensory tentacles fringe the mantle of the Giant scallop (*Pecten maximus*). The eyes detect sudden changes in light caused by arrival of would-be predators. In the free-swimming scallops the mantle margin is also used to direct jets of water propelling the mollusk to safety.

▶ **Rows of rasping teeth** on radula of a herbivorous top shell (*Trochus* species). Carnivorous mollusks have fewer, larger, more pointed teeth on a narrower radula.

◀▼ **Iridescence** LEFT inside a Pearly nautilus shell is caused by reflected light being refracted as it passes calcite back out of the shell through calcite crystals. The mother-of-pearl inside nautilus, top, turban, ormer and other shells, and the pearls formed in the mantle cavity of oysters and mussels are made of the same material. Four shell layers are now recognized BELOW: (1) an inner nacreous layer of calcite crystals in flat layers parallel to the surface; (2) a cross-lamellar component of oblique crystals; (3) a chalky, often pigmented, prismatic layer, usually the thickest, of crystals set at right angles to the surface; and (4) a shiny, horny outer layer, the periostracum, of proteins and polysaccharides.

ink along the outside of the lip. After a few days new shell will be seen in front of the ink mark. Newly secreted shell is thin, but gradually attains the same thickness as the rest. Although, in the event of damage, a repair can be made further back from the mantle edge, it will be a rough patch and not contain all the layers of normal shell. When growth stops for a while during cold weather, in drought, or a time of starvation, a line forms on the shell which continues to be visible after resumption of growth. A number of mollusks (eg cockles) normally have regular marks recording interruptions of their growth.

The cross-lamellar component of the shell, revealed by high magnification, consists of different layers of oblique crystals, each layer with a different orientation, rather like the structure of radial car tires This is thought to give greater strength without extra weight or bulk.

Shell is mostly composed of calcium carbonate, in calcite form (in the prismatic layer) and in argonite form (in the cross-lamellar layers), together with some sodium phosphate and magnesium carbonate. The mineral component of the shell is laid down in organic crystalline bodies in a matrix of fibrous protein and polysaccharides (conchiolin) secreted by the mantle. Snails can store calcium salts in cells of the digestive gland (hepatopancreas). When needed for growth or shell repair the salts are transported to the mantle by migratory cells.

Molluskan shells show great variation in shape, size, thickness, sculpture, surface texture and shine. Marine examples are often thick and heavy, while land snails, lacking the support of water, tend to have thinner shells.

The spirally coiled shells of gastropods range from tall and spindle-shaped to flat and disk-like; the body whorl containing the animal itself may be small, or enlarged to occupy most of the shell; and likewise the aperture or mouth of the shell can be open or constricted and armed with a range of teeth or ribs – in whelks and other carnivorous sea snails there is a groove (siphonal canal) to house the siphon. With the shell apex uppermost, the mouth of the shell in most mollusks is on the right-hand (dextral) side, but some species normally have the mouth on the left (sinistral). Some genera, such as Hawaiian tree snails and *Amphidromus* (both tropical land snails), and the temperate-zone whorl snails, have both dextral and sinistral examples. In some normally dextral species an occasional sinistral species may be found, but this is unusual.

scratch marks on the surface of rock.

In the carnivorous cone shells each tooth is separated from the membrane and is a harpoon-like structure which is delivered into the body of the prey (often a fish or a worm) to facilitate penetration of an accompanying nerve poison. The tiny sacoglossan sea slugs feed on thread-like algae—their radula teeth are adapted to pierce individual algal cells.

The shell

Mollusks usually hatch from the egg complete with a tiny shell (protoconch) that is often retained at the apex of the adult shell. This calcareous shell, into which the animal can withdraw, is often regarded as a hallmark of a mollusk. For the living mollusk, the shell provides protection from predators and mechanical damage, while on land and on the shore it helps to prevent loss of body water. Empty shells have long been a source of fascination in themselves, and many people collect them.

New growth occurs at the shell lip in gastropods and along the lower or ventral margin in bivalves. Shell is secreted by glandular cells, particularly along the edge of the mantle. This is easy to demonstrate in young land snails, by painting waterproof

THE 7 CLASSES OF MOLLUSKS

Includes species, genera and families mentioned in the text. For reasons of space, divisions such as suborders and superfamilies, important in some groups, are omitted. The sequence of families reflects relationships.

Monoplacophorans
Class: Monoplacophora

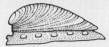

Five species of deep-sea segmented limpets, eg *Neopilina galathea*.

Solenogasters, chaetoderms
Class: Aplacophora

About 200 worm-like marine species in subclasses Solenogastres (eg *Epimenia verrucosa*) and Caudofoveata.

Chitons or coat-of-mail shells
Class: Polyplacophora

About 500 species including the **Giant Pacific chiton** (*Amicula stelleri*) and *Mopalia* species.

Slugs and snails (gastropods)
Class: Gastropoda

About 60,000–75,000 species in 3 subclasses.

Prosobranchs or operculates
Subclass: Prosobranchia

Order Diotocardia
Slit shells (Pleurotomariidae), eg *Pleurotomaria* species. **Ormers** and **abalones** (Haliotidae), eg *Haliotis*, *Ormer* species. **Slit and keyhole limpets** (Fissurellidae), eg *Diodora*, *Fissurella* species, **Great keyhole limpet** (*Megathura crenulata*). **True limpets** (Patellidae), eg **Common limpet** (*Patella vulgata*), **Blue-rayed limpet** (*Patina pellucida*). **Top shells** (Trochidae), eg **Thick top shell** (*Monodonta lineata*). **Turban shells** (Turbinidae), eg **Tapestry turban** (*Turbo petholatus*). **Pheasant shells** (Phasianellidae), eg **Pheasant shell** (*Tricolia pullus*), **Australian pheasant shell** (*Phasianella australis*).

Order Monotocardia
Mesogastropods
Apple snails (Ampullariidae), eg *Pila*, *Pomacea* species. **River snails** (Viviparidae), eg *Viviparus viviparus*. **Winkles** or **periwinkles** (Littorinidae), eg **Dwarf winkle** (*Littorina neritoides*), **Flat winkle** (*L. littoralis*), **Edible winkle** (*L. littorea*). **Round-mouthed snails** (Pomatiidae), eg *Pomatias elegans*. **Spire snails** (Hydrobiidae), eg **Jenkins' spire shell** (*Potamopyrgus jenkinsi*). Family Omalogyridae, eg *Ammonicera rota*. **Vermetids** (Vermetidae), eg *Vermetus* species. **Sea snails** (Janthinidae), eg **Violet sea snail** (*Janthina janthina*). Family Styliferidae (parasites), eg *Stylifer*, *Gasterosiphon* species. Family Eulimidae (parasites), eg *Eulima*, *Balcis* species. Family Entoconchidae (parasites), eg *Entoconcha*, *Entocolax*, *Enteroxenos* species. **Slipper limpets**, **cup-and-saucer** and **hat shells** (Calyptraeidae), eg **Atlantic slipper limpet** (*Crepidula fornicata*). Family Capulidae (parasites), eg *Thyca* species. **Ostrich foot shells** (Struthiolariidae), eg *Struthiolaria* species. **Conch shells** (Strombidae), eg **Pink conch shell** (*Strombus gigas*). **Cowries** (Cypraeidae), eg **Money cowrie** (*Cypraea moneta*), **Gold ringer** (*C. annulus*). **Necklace** and **moon shells** (Naticidae), eg *Natica* species.

Neogastropods
Whelks (Buccinidae), eg **Edible whelk** (*Buccinum undatum*). **Dog whelks** (Nassariidae), eg *Bullia tahitensis*. **Spindle shells** (Fasciolariidae), eg *Fasciolaria* species. **Rock shells** or **murexes** (Muricidae), eg *Murex* species, **Common dog whelk** (*Nucella lapillus*), **Oyster drill** (*Urosalpinx cinerea*). **Volutes** (Volutidae), eg *Voluta* species. **Olives** (Olividae), eg *Oliva* species. **Turret shells** (Turridae). **Cones** (Conidae), eg **Courtly cone** (*Conus aulicus*), **Geographer cone** (*C. geographicus*), **Marbled cone** (*C. marmoreus*), **Textile cone** (*C. textile*), **Tulip cone** (*C. tulipa*). **Auger shells** (Terebridae), eg *Terebra* species.

Sea slugs and bubble shells
Subclass: Opisthobranchia

Order Bullomorpha—bubble shells
Acteon shells (Acteonidae), *Acteon* species. **Bubble shells** (Hydatinidae), eg *Hydatina* species. **Cylindrical bubble shells** (Retusidae), eg *Retusa* species. **Bubble shells** (Bullidae), eg *Bullaria* species. **Lobe shells** (Philinidae) eg *Philine* species. **Canoe shells** (Scaphandridae), eg *Scaphander* species.

Order Pyramidellomorpha
Pyramid shells (Pyramidellidae).

Order Thecosomata
Sea butterflies or **pteropods** (Spiratellidae), eg *Limacina* species.

Order Gymnosomata
Sea butterflies or **pteropods** (Clionidae), eg *Clione* species.

Order Aplysiomorpha
Sea hares (Aplysiidae), eg *Aplysia* species.

Order Pleurobranchomorpha

Order Acochlidiacea
Family Hedylopsidae, eg *Hedylopsis* species. Family Microhedylidae, eg *Microhedyle* species.

Order Sacoglossa
Bivalve gastropods (Julidae), eg *Berthelinia limax*. Family Elysiidae, eg *Elysia*, *Tridachia* species. Family Stiligeridae, eg *Hermaea* species. Family Limapontiidae, eg *Limapontia* species.

Order Nudibranchia—shell-less sea slugs
Sea slugs (Dendronotidae), eg *Dendronotus* species. **Sea lemons** (suborder Doridacea). Family Tethyidae, eg *Melibe leonina*. Family Aeolidiidae, eg **Common gray sea slug** (*Aeolidia papillosa*). **Floating sea slugs** (Glaucidae), eg *Glaucus atlanticus*, *G. marginata*.

Lung-breathers or pulmonates
Subclass: Pulmonata

Order Systellommatophora—tropical slugs
Eg *Veronicella* species.

Order Basommatophora—pond and marsh snails
Operculate pulmonates (Amphibolidae), eg *Salinator* species. **Dwarf pond snails** (Lymnaeidae), eg *Lymnaea trunculata* **Bladder snails** (Physidae), eg *Physa* species, **Moss bladder snail** (*Aplexa hypnorum*). **Ramshorn snails** (Planorbiidae), eg *Bulinus*. **Freshwater limpets** (Ancylidae).

Order Stylommatophora—land snails and slugs
Hawaiian tree snails (Achatinellidae), eg *Achatinella* species. **Whorl snails** (Vertiginidae), eg *Vertigo* species. **African land snails** (Achatinidae), eg *Archachatina marginata*. Family Oleacinidae, eg *Euglandina* species. **Shelled slugs** (Testacellidae), eg *Testacella* species. **Glass snails** (Zonitidae), eg **Garlic snail** (*Oxychilus alliarius*), *Aegopis verticillus*. Family

Limacidae, eg **Gray field slug** (*Deroceras reticulatum*), **Great gray slug** (*Limax maximus*). Family Chamaemidae, eg *Amphidromus* species. Family Helicidae, eg **Brown-lipped snail** (*Cepaea nemoralis*), **Common garden snail** (*Helix lucorum*), **Roman snail** (*H. aspersa*), **Desert snail** (*Eremina desertorum*), *Eobania vermiculata*, *Otala lactea*. **Carnivorous snails** (Streptaxidae).

Tusk or tooth shells
Class Scaphopoda

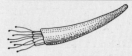

About 350 species of marine sand burrowers.
Family Dentaliidae, eg **Elephant tusk shell** (*Dentalium elephantinum*). Family Cadulidae, eg *Cadulus* species.

Clams and mussels (bivalves)
Class: Bivalvia (or Pelecypoda)

About 15,000–20,000 species of sea, brackish and freshwater.

Order Protobranchia
Includes **nut shells** (Nuculidae), eg *Nucula* species.

Order Taxodonta—ark shells, dog cockles
Includes **dog cockles** (Glycimeridae), eg *Glycimeris* species.

Order Anisomyaria
Mussels (Mytilidae), eg **Date mussels** (*Lithophaga* species), **Edible mussel** (*Mytilus edulis*), *Botulus*, *Fungiacava* species. **Pearl oysters** (Pteriidae), eg *Pinctada martensii*. **Scallops** (Pectinidae), eg *Pecten* species, **Queen scallop** (*Aequipecten opercularis*). **File shells** (Limidae), eg *Lima* species. **Saddle oysters** (Anomiidae), eg **Window oyster** (*Placuna placenta*). **Oysters** (Ostreidae), eg *Ostrea*, *Crassostrea* species.

Order Schizodonta
Freshwater pearl mussel (Margaritiferidae), *Margaritifera margaritifera*. **Freshwater** or **river mussels** (Unionidae), eg *Unio*, *Anodonta*, *Lampsilis* species. **Tropical freshwater mussels** (Aetheriidae), eg *Aetheria* species.

Order Heterodonta
Cockles (Cardiidae). **Giant clams** (Tridacnidae), eg *Tridacna gigas*,

T. crocea. **Pea shells** (Sphaeriidae), *Pisidium* species. **Fingernail clam** (Corbiculidae), *Corbicula manilensis.* **Atlantic hard-shell clam** (Arctidae), *Arctica islandica.* **Zebra mussel** (Dreissenidae), *Dreissena polymorpha.* **Ruddy lasaea** (Erycinidae), *Lasaea rubra.* Family Galleomatidae, eg *Devonia perrieri.* **Montagu shells** (Montacutidae), eg *Montacuta* species, *Mysella bidentata.* **Venus** and **carpet shells** (Veneriidae), include **Venus shells** (*Venus* species), eg **quahog** or **Hard-shell clam** (*V. mercenaria*), **smooth Venus** (*Callista* species), **carpet shells** (*Venerupis* species). **False piddock** (Petricolidae), *Petricola pholadiformis*, **oval piddocks** (*Zirfaea* species). **Wedge shells** or **bean clams** (Donacidae), eg *Donax* species. **Tellins** (Tellinidae), eg *Tellina folinacea.*

Order Adepedonta
Trough shells (Mactridae), eg *Spisula* species. **Razor shells** (Solenidae), eg *Ensis* species. **Gapers** (Myidae), eg *Mya, Platyodon* species. **Rock borer** or **Red nose** (Hiatellidae), (*Hiatella arctica.* **Flask shells** (Gastrochaenidae), eg *Gastrochaena* species. **Piddocks** (Pholadidae), eg *Pholas* species, **wood piddocks** (*Xylophaga* species), **shipworms** (Teredinidae), eg *Teredo* species.

Order Anomalodesmata —septibranchs
Dipper clams (Cuspidariidae), eg *Cuspidaria* species.

Cephalopods
Class: Cephalopoda

About 650 marine, mostly pelagic, species.

Nautiloids
Subclass: Nautiloidea
Eg **Pearly nautilus**, *Nautilus* species

Ammonites (extinct)
Subclass: Ammonoidea
Eg **Giant ammonite** (*Titanites titan*).

Subclass: Coleoidea

Order Decapoda—cuttlefishes
Cuttlefishes (eg *Sepia, Sepiola* species), **squids** (eg *Loligo* species), **flying squid** (*Onycoteuthis* species), **Giant squid** (*Architeuthis harveyi*); also extinct **belemites.**

Order Vampyromorpha
Eg *Vampyroteuthis infernalis.*

Order Octopoda—octopuses
Eg *Octopus* species, **Paper nautilus** (*Argonauta* species).

The nautilus, an exception among living cephalopods, has a light, brittle, spiral shell. In section this is seen to be divided, behind the outermost body chamber, by thin walls into progressively smaller earlier body chambers. Each wall (septum) has a central perforation which in the living animal is traversed by a thread-like extension of the body, the siphuncle, which extends to the shell apex. The pressure of gas in the chambers affects buoyancy.

Many shells are strongly sculpted into ribs, lines, beading, knobs or spectacular spines. Such detail, much admired by shell collectors, is also used in the identification of species. The surface of the shells is sometimes rough, but in certain examples, such as cowries and olive shells, it may be smooth and glossy. Many tropical shells are very colorful and may also have attractive patterns and markings.

A number of mollusks in different groups have a reduced shell or none at all. Some sea slugs retain external shells, others have thin internal shells and some (the nudibranchs) like land slugs are without any shell. Shell-less sea slugs have evolved to swim as well as crawl, to squeeze into small crannies, and to develop secretions and body color as means of defense. An external shell is also lacking in some parasitic gastropods, in the worm-like aplacophorans and in most cephalopods.

The evolution of mollusks
The success of the mollusks has been due to their adaptability of structure, function and behavior. Mollusks are thought to be derived from either an ancestor of the Platyhelminthes (planarians, flukes, tapeworms) or from the arthropod annelid line, the latter having trochophore larvae, as do most mollusks.

There is fossil evidence of mollusks in some of the oldest rocks bearing fossils, dating back over 530 million years to the Cambrian period. Mollusks soon evolved in different directions and the modern classes had largely separated out by the end of the Cambrian, 500 million years ago. The early fossils were all marine. Land snails appeared in the coal-measure forests of 300 million years ago, but land snail fossils are rare until deposits of the Tertiary (65–2 million years ago).

The original molluskan shell was probably cap-like. Many families, such as limpets, have reverted to that form from the spiral coiling that was widespread in gastropods (and still is) and in the few remaining shelled cephalopods. Nautiloids

(now represented by only six living species) gave rise to some 3,000 known fossil species, many of which flourished in the Paleozoic seas, but they dwindled in the Mesozoic (225–65 million years ago) when ammonites expanded. After a successful period, ammonites suddenly vanished in their turn at the end of the Cretaceous, along with the dinosaurs.

There is no central theme to molluskan evolution. Different groups of mollusks have adapted to similar habitats, often adopting similar characteristics (convergent evolution). Among bivalves, for example, both true piddocks and the False piddock bore into mudstone but, although the outsides of their shells are similar, the latter is more closely related to venus shells and its shell-hinge teeth are quite different. The bivalve shell of members of the class Bivalvia even has its counterpart in a different class, the bivalve gastropods (see p127).

Isolated islands as in the Pacific show a high level of speciation (evolution of new species) and forms that are endemic (limited to that island), because of the separation of the snail population from other larger populations. In Hawaii there are even local color forms of tree snails in isolated valleys.

Respiration and circulation
Mollusks originated in the sea—and their basic method of breathing is by gills which extract dissolved oxygen from the surrounding water. The typical molluskan gill (ctenidium) consists of a central axis from which rows of gill filaments project on either side (bipectinate). Blood vessels enter the filaments and the surface of each filament is covered with cells bearing cilia; some of these cells create a current in the water with their long cilia, and others pick up food particles. An inhalant current of water enters the gill on one side, passes between the filaments and goes out as an exhalant current on the other side.

In some mollusks the gill serves only for respiration, while in others (eg most bivalves) the two enlarged gills have a dual role of feeding as well as respiration. Gills are delicate structures, which need to be kept clear of clogging particles: the ciliary devices evolved for cleaning the gills later became adapted for feeding. The mantle is often developed at the rear end into a tube (siphon) which projects and takes in water, testing it with sensory cells and tentacles on the way; in some bivalves (eg the tellinids) there is a separate siphon for the exhalant current.

Blood is pumped around the molluskan

body by a heart, and is distributed to the tissues by arteries but, in all except cephalopods, it has to make a slow passage back through blood spaces (hemocoel).

Monoplacophorans and chitons have several pairs of gills in the mantle cavity. In prosobranch or operculate gastropods there is typically one pair of gills, but in the more highly evolved groups the gill is reduced and lost on one side, so winkles and whelks have only one gill in the mantle cavity. The more primitive prosobranchs (eg slit shells, ormers, slit and keyhole limpets) still have two gills. These limpets have lost the typical molluskan ctenidium, replacing it by numerous secondary gills of different structure which hang down around the mantle cavity. Most prosobranchs are aquatic, breathing by gills, although some have adapted to life on land and breathe by a vascularized mantle cavity or lung.

During drought, land prosobranchs can spend long periods of time inactive inside the shell, sealed off by an operculum. Some tropical land species of the family Annulariidae have developed a small tube of shell material behind the operculum which enables the snail to obtain air when the operculum is enclosed. The tropical apple snails live in stagnant water, and these have long siphons which reach above the surface of water to breathe air.

Among the sea slugs and bubble shells, the primitive shelled species such as *Acteon* and the sea hares have a gill in the mantle cavity, but in the shell-less forms there are either secondary gills, as in the sea lemons, or respiration takes place directly through the skin, which may have its surface area increased by numerous papillae.

The pulmonates (land snails and slugs, pond snails), as the name implies, are essentially lung breathers, having lost the gill, and breathe air from a highly vascularized mantle wall. The finely divided blood vessels of the respiratory surface can often be seen as silvery lines through the thin shell. The entrance to the mantle cavity is sealed off and opens by a breathing pore (pneumostome) whenever an exchange of air is needed. The pulmonate pond snails usually breathe air and come up to the surface to open the breathing pore above water.

Some deepwater pond snails of lakes have reverted to filling the mantle cavity with water and no longer use air. While freshwater prosobranchs breathe by gills and are more often found in oxygenated waters of rivers and streams, the pulmonates, breathing air, are better adapted for living in the still stagnant water of ponds and ditches.

▲ **An active predator,** the Common cuttlefish (*Sepia officinalis*) has grabbed a Common prawn (*Palaemon serratus*) by shooting out two long arms. These draw the prey into the grasp of the eight short arms which secure it so that the cuttlefish's sharp beak-like jaws can get to work on the prawn's shell.

◄ **One-way system.** In bivalves such as the Edible mussel, water enters the mantle cavity by an inhalant siphon (the one with the frilly tentacles). After passing over the gills, where food is filtered off as well as respiratory gases exchanged, the water is exhaled by the smaller siphon.

► **Naked sea slugs** (*Polycera quadrilineata*) feeding on a sea mat (*Electra* species). The sea slugs and bubble shells (opisthobranchs) include herbivores, suspension feeders, specialist carnivores and parasites, but the majority are carnivores feeding on encrusting marine animals.

Bivalves are all aquatic and breathe by gills. In primitive families, including the nut shells, the gills are relatively small, being only used for breathing, not feeding.

Cephalopods breathe by gills in the mantle cavity. The system is efficient, for there is a greater flow of water through the mantle cavity due to its use in jet propulsion, and faster blood circulation through a closed blood system with veins as well as arteries. The branchial hearts of most cephalopods are not present in nautiluses,

---

Cones—The Venomous Snails

The 400–500 species of cone shells are found mostly in tropical and subtropical waters. These great favorites among collectors are unusual exceptions among mollusks, being directly harmful to humans.

Cones are carnivores, taking a range of prey, from marine worms to sizeable fish, and those feeding on fish are most dangerous. In common with other carnivorous gastropods, such as whelks, cones have a proboscis. This muscular retractable extension of the gut carries the mouth, radula and salivary gland forward to reach food in confined spaces or for other reasons at a point distinct from the animal. When the probing, extended proboscis of a cone touches a fish, it embeds one of the harpoon-like teeth on the tip of the radula into the prey, accompanied by a nerve poison which paralyzes the fish; the cone swallows the fish whole.

The Geographer cone, Textile cone and Tulip cone have been known to kill humans, while others like the Courtly cone, Marble cone and *Conus striatus* can cause an unpleasant, although not fatal, sting.

---

which rely instead on duplication of gills to meet the respiratory needs.

Feeding

The primitive mollusk probably fed on small particles, the macrophagous habit (eating large particles) developing later. Most mollusks, except bivalves, feed using the radula (see above). The bivalves and some gastropods (eg slipper limpets of the USA, ostrich foot of New Zealand, and river snails of Europe) are ciliary feeders, either straining food from seawater (filter feeding) or sucking in sludge off the bottom (deposit feeding).

Another typical molluskan feature, found in some prosobranchs and bivalves, is the stomach, its wall protected by chitinous plates from a pointed, forward-projecting style which winds round and brings the string of food from the esophagus into the stomach. The style also secretes digestive enzymes.

The gastropods include browsing herbivores, ciliary feeders, detritus feeders, carnivores and parasites. The muscular mouthparts (buccal mass) and gut show adaptions reflecting this variety. Limpets, top shells and winkles browse on algae and other encrustations on rocks, while the Flat periwinkle eats brown seaweed and the Blue-rayed limpet rasps at the fronds and stems of oarweed (*Laminaria* species). Some will eat carrion while others attack live animals. The whelks, murexes or rock shells, volutes, olives, cones, auger and turret shells (neogastropods) are specialized for a carnivorous diet—the shell has a siphonal canal housing a siphon which directs water over taste cells in the osphradium (a chemoreceptor in the mantle cavity), which helps in the detection of food. Certain carnivorous gastropods, like the Dog whelk and the necklace shells, drill holes in the shells of other mollusks which are then consumed. The murex or rock shells bore mechanically, but necklace shells use acid to soften the shell before excavating with the radula.

The sea slugs and bubble shells include herbivores, suspension feeders, carnivores and parasites, but the majority are carnivores feeding on encrusting marine animals. The sea slug *Melibe leonina*, from the west coast of the USA, is an active swimmer and adapted to feeding on crustaceans, which it catches with the aid of a large cephalic hood: the radula is absent. Shelled opisthobranchs, like species of canoe shell, lobe shell and cylindrical bubble shell, feed on animals in the sand, including mollusks. The inside of the gizzard is lined with special

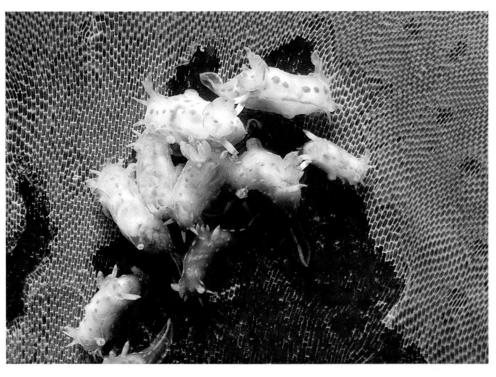

plates which they use to crush the prey. The curious pyramidellids with small coiled shells are parasitic in a range of marine animals.

Pulmonates are chiefly herbivorous, although many of them feed on dead rather than living plant material, and pond snails often consume detritus and mud on the bottom. In these animals the radula is broad, with large numbers of small teeth. There is no style nor chitinous gizzard plates as found in bubble shells. The smaller land snails retain a microphagous diet, while a few species from different families have become carnivorous. These include the shelled slugs that eat earthworms, and a number of tropical land snails, such as the family Streptaxidae and genus *Euglandina*, which eat other snails. Some glass snails (family Zonitidae) have carnivorous tendencies and the large glass snail *Aegopis verticillus* of eastern Europe readily eats land snails.

The more primitive bivalves feed on detritus, which is pushed into the mouth by labial palps, but most modern bivalves are ciliary feeders, making use of phytoplankton, while others take in detritus from the surface of the substrate with siphons. In the more adanced bivalves the gills are used for filtering food and conveying it to the mouth by wrapping it in mucus and passing it along food grooves to the mouth. The crystalline style and stomach plates are well developed in bivalves. The woodboring shipworms have cellulase enzymes used to digest wood shavings. The curious group of bivalve septibranchs have lost the gill and have reduced labial palps and style. They are scavengers, sucking juices of dead animals. Species of the bivalve genus *Entovalva* are parasites inside sea cucumbers, and some of the freshwater mussels are parasitic in fish in the early stages of their life history.

The carnivorous cephalopods mostly catch fish, although the slower-moving octopus takes crustaceans. The radula is relatively small, but the prey is seized by jaws with a hard beak. In cephalopods, enzymes are secreted by gland cells into the tubules of the digestive gland where extracellular digestion takes place: this contrasts with the ingestion of food particles by cells of the digestive gland (intracellular digestion) in other mollusks. Extracellular digestion is a feature in which cephalopod body organization is in advance of the rest of the mollusks and parallels the situation in vertebrates.

Excretion

In mollusks there is a kidney next to the heart which extracts nitrogenous waste from the blood. The excretory duct runs alongside the rectum to the pore at the mantle edge in pulmonates, but in prosobranchs there is a simple opening on the side of the kidney directly into the mantle cavity. In some mollusks certain minerals are selectively resorbed. There is little water regulation in marine mollusks but considerable activity in those of freshwater. Land mollusks conserve their water and little goes out with the excreta, nitrogenous waste being in a insoluble crystalline form and often stored in the kidney. Bivalves give off their nitrogenous waste as ammonia or its derivatives. In some opisthobranchs excretory waste is discharged into the gut.

Breeding

Eggs of mollusks vary considerably. Some are shed into water before fertilization as in bivalves. Many mollusk eggs are very small but those of cephalopods are large and yolky. When fertilization is internal, elaborate egg cases may be secreted and very often gastropods lay eggs in large clutches. Some winkles and water snails deposit eggs in a jelly-like matrix often attached to vegetation. Necklace shells form stiff collars of egg cases which are large for the size of the mollusk, while whelks and murexes deposit eggs in leathery capsules attached to rocks and weed. Land snail eggs tend to be buried in soil: some are contained in a transparent envelope but others have a limy eggshell, and the eggs of one of the large African land snails, *Archachatina marginata*, are of the size and appearance of a small bird's egg. Egg masses of sea slugs can be quite spectacular when found in rock pools.

In aquatic mollusks there is usually a planktonic larva, the primitive trochophore, of short duration, and/or the characteristic veliger larva that develops from it (often within the egg capsule), with shell and ciliary lobes. Some retain the egg inside the body of the female, or in the capsule, from which the young emerge as miniature adults. The veliger lives in the plankton, feeding on algae for a day to several months before settling. Vast numbers of molluskan larvae occur in the plankton and many of them are eaten by other members of the zooplankton or by filterfeeders, or perish when unable to find a suitable habitat to settle.

In land and freshwater prosobranchs the veliger stage is suppressed, and a snail hatches direct from the egg. Pulmonates also lack the veliger stage, and floating larvae occur in only some freshwater bivalves. The pelagic larva was important

▶ **Mating in a shoal,** squid (*Loligo* species) off Catalina Island, California. Male squids and other male cephalopods parcel up their sperm into a packet (spermatophore) carried on a specially modified arm called a hectocotylus. This is used to transfer the spermatophore to the mantle cavity of the female, where fertilization occurs. The hectocotylus may be broken off there.

Gastropod snails and slugs more often copulate. This safe transfer of the male sperm by internal fertilization was a prerequisite for gastropods' unique success story among mollusks—the colonization of land.

Some prosobranch snails, such as Jenkins' spire shell, are found in all-female populations that breed parthenogenetically, without the need of males. This has enabled Jenkins' spire shell to colonize streams and ponds over a large part of northwestern Europe within a century.

Other mollusks—monoplacophorans, chitons, bivalves and some gastropods—fertilize eggs externally: both female and male gametes are shed into the sea, or male gametes are brought into the female's mantle cavity on the inhalent current.

▼ **Seen from above, a floating sea slug,** *Glaucus atlanticus*, feeds on a Portuguese man-of-war, a colony of specialized hydrozoan polyps, in the surface waters of the ocean.

in establishing the freshwater Zebra mussel that first came to Britain in the first half of the 19th century. Freshwater or river mussels brood the eggs in the gills and release them as glochidia larvae parasitic on fish.

It is in the sea snails and limpets (prosobranchs) and sea slugs and bubble shells (opisthobranchs) that the veliger is most varied. Chitons and more primitive prosobranchs such as slit shells and ormers have a trochophore larva, with a horizontal band of ciliated cells, which only lasts for a few days. Mollusks with veligers in the plankton for several weeks have a better opportunity for dispersal. At metamorphosis the ciliated lobes (velum) by which the veliger swims and feeds are engulfed, the mollusk ends its planktonic life and sinks to the bottom. Bivalve veliger larvae also occur in the plankton but in some, such as the small midshore *Lasaea rubra*, the young hatch as bivalves and establish themselves near the parent colony. After the bivalve veliger stage is an intermediate pediveliger when the larva searches for a suitable place to settle; if none is found, the velum can be reinflated and the larva is carried to other sites.

Nervous system and sense organs

The brain and eye of the cephalopod are the most highly developed of any invertebrate. The molluskan nervous system essentially consists of pairs of ganglia (masses of nerve tissue), each ganglion linked by nerve fibers. In the more primitive groups the individual ganglia are well separated, but in more highly evolved mollusks, such as land snails and whelks, there is both a shortening of the connectives, bringing the ganglia into closer association, and a concentration of most of the ganglia in the head. The ring of ganglia round the front part of the gut (esophagus) in mollusks, compares with the nervous system of annelids and arthropods. The main pairs of ganglia in mollusks are the cerebral, pleural, pedal, parietal, visceral and buccal (receiving impulses from the head, mantle, foot, body wall and internal organs), and there is a pedal "ladder" arrangement in monoplacophorans, chitons and the more primitive prosobranchs such as slit shells. In prosobranch gastropods there is the further complication of torsion (see p126).

Most mollusks are sensitive to light, which can be detected by sensors in the shell plates of chitons, the black eyespots associated with the tentacles of most gastropods, and the very elaborate cephalopod eye. The balance of the animal is maintained by special sense organs called statocysts. Prosobranchs have a special chemosensory organ, the osphradium, but there are less specialized patches of chemonsensory cells in other mollusks. The terrestrial slugs have a sense of smell used to find food.

Mollusks are also sensitive to touch: the suckers of octopuses can discriminate texture and pattern (see box p134), while the lower pair of tentacles of land pulmonates are largely tactile and function in feeling the way ahead.

Movement in mollusks

Some mollusks (eg mussels and oysters) anchor themselves to one place, but most move around in pursuit of food, for mating and to escape enemies. The octopus crawls, using suckers on its arms, modifications of the foot. Despite the mollusks' slow image, some, like cephalopods, can swim surprisingly fast.

Usually the foot is the organ involved in locomotion, which involves gliding over the surface of seabed, rock or plant. Land snails and slugs, particularly, lay down a lubricating and protective film of slime or mucus— the silvery trails seen on garden paths and walls. Some species "leap" with long stretches of the foot; the head lunges forward and attaches to ground ahead. Lobes of the foot (parapodia) are often developed for swimming in the sea slugs, and the pelagic thin-shelled *Limacina* and slug-like *Clione* species also have swimming "wings." The sea slugs, freed from the restrictions of a shell, can swim by lateral movements using muscles in the body wall.

Planktonic larvae and some small gastropods move primarily by the beating action of cilia. Veliger larvae of gastropods and bivalves have minute, hair-like cilia on the lobes of the enlarged velum which are used for feeding, respiration and locomotion. They are also able to adjust their depth by retreating into the shell to sink, then reexpanding the velum to halt the descent. Many pond snails move by cilia on the sole of the foot and they can glide along the surface film of water by this method. On land, where the body weight is not supported by water, snails moving by ciliary means are more likely to be the smaller species. Bivalves may swim by shell-flapping (scallops), they may "leap" across the surface (cockles), or burrow. Cockles use the pointed foot for moving the shell across the surface of the sand. Most bivalves, however, use the foot (which can be of considerable size) for burrowing: it is pushed forward and expanded by blood entering the pedal hemocoel

► **Jet swimming** by a scallop (*Pecten maximus*), as it escapes from a starfish (here *Marthasterias glacialis*). In flapping the two valves of its shell, the scallop expels jets of water from its mantle cavity. Cockles use their pointed foot for "leaping" across the surface of the sand, but most bivalves use the foot (which may be of considerable size) for burrowing.

▼ **A helmet shell** (*Phalium labiatum*) on the move. On top of the large, muscular foot, at the rear, can be seen the lid-like operculum that protects the animal when it withdraws into its shell. The siphon, and tentacles with eyes at their bases, can also be seen.

As in most mollusks, the muscular foot provides locomotion, by means of wave-like movements that in this case are longitudinal, as in winkles and top shells, for example, while in the terrestrial pulmonates the muscle waves are transverse, traveling from head to tail or the reverse.

(blood space); the muscles of the foot also contract and it then changes shape, often forming an anchor. Further contraction brings the shell down into the sand; as the shell closes, water jetted out of the mantle cavity can help to loosen the sand ahead. The digging cycle then starts again.

Solenogasters and chaetoderms

The shell-less aplacophorans are curious worm-like creatures found in the mud of marine deposits, usually offshore. This small and little-understood group was once classified with the chitons, but is now placed in a group of its own. Indeed, recent research suggests that the class Aplacophora should be divided into two classes, the solenogasters (class Solenogastres) and the chaetoderms (class Caudofoveata).

Aplacophorans do not have an external shell, although there may be tiny, pointed calcareous spicules in the skin, sometimes of a silvery, "fur-like" appearance. In common with other mollusks, most possess a radula, a mantle, mantle cavity, a foot and a molluskan-type pelagic larva. There are dorso-ventral muscles crossing the body, which are reminiscent of similar structures in flatworms, flukes and tapeworms.

Most of the 200 or so species are small, but the solenogaster *Epimenia verrucosa* can reach 12in (30cm) in length. Solenogasters are fairly mobile and can twist their bodies around other objects; the foot is reduced to a ventral groove. Chaetoderms (named for their spiny skin), live in mud, moving rather like earthworms but spending much of their time in a burrow. The radula is often reduced in this group.

Monoplacophorans

First of the seven classes within the phylum, the Monoplacophora is a small group of primitive mollusks that were originally thought to have become extinct about 400 million years ago. The flattish cap-like shell of monoplacophorans resembles that of limpets, and the groups used to be classified with the gastropods.

In 1952 living monoplacophorans were collected 11,700ft (3,750m) down in a Pacific Ocean deep-sea trench off South America, and a new species of monoplacophoran, *Neopilina galathea*, was described. The shell is pale, fairly thin, cap-like in shape and about 1in (2.5cm) long. On the inner surface of the shell, instead of the single horseshoe-shaped muscle scar of limpets, there are several pairs of muscle scars in a row on either side.

The particularly interesting feature of

Neopilina galathea is the repetition of pairs of body organs: there are eight pairs of retractor muscles attaching the animal to the shell, 5–6 pairs of gills, 6–7 pairs of excretory organs, a primitive ladder-like pedal-nervous system with 10 connectives across, and two pairs of gonads. Although there is a parallel in chitons (see below), in monoplacophorans this repetition is taken much further. In consequence, it has been suggested that mollusks evolved from an annelid/arthropod ancestor, rather than from an unsegmented flatworm, and that the original segmentation of the body was lost during early molluskan evolution. Some other researchers disagree, considering the repetition of body organs to be a more recent, secondary character, rather than a primitive one.

In *Neopilina galathea* there is a molluskan radula and posterior mantle cavity and anus, showing that torsion (the twisting of body organs found in gastropods) did not occur in the monoplacophorans. Since the original discovery, further living species of monoplacophorans have been found in deepwater trenches in other parts of the world such as Aden, and now five living species are described.

Chitons

Chitons, or coat-of-mail shells, have a distinctive oval shell consisting of eight plates bounded by a girdle. They are exclusively marine and, with the exception of a few deepwater species, mostly limited to shores and continental shelves. Beneath the shell with its low profile and stable shape, the animal attaches to the rock by a sucker-like foot. The plates of the shell are well articulated—chitons can roll up into a ball when disturbed. The articulations are also an advantage when moving over the uneven surface of rocks.

Although there are no obvious eyes, chitons are sensitive to light through light

receptors in the shell. Chitons are found mostly on rocky shores. When a boulder is turned over, chitons on the underside quickly move down again out of the light.

Chitons have remained substantially unchanged since the Cambrian period 600–500 million years ago. The different species are identified by the relative width of the girdle protecting the mantle, by the sculpturing of the shell valves and the bristles they bear, and also the teeth and the surfaces of the joins between the valves. Most chitons are 0.4–1.2in (1–3cm) long, but some, like the Giant Pacific chiton can reach 8–12in (20–30cm). The larger and more spectacular species are found on the Pacific coast of the USA and off Australia.

Chitons are browsers, rasping algae and other encrusting organisms off the rock with the hard teeth of the radula. One family, the Mopaliidae, is carnivorous, feeding on crustaceans and worms.

The anatomy of the chiton is closer to that

▲ **Delicate colors and patterns** of the mantle that envelops the shell of this Australian cowrie bear little relation to the colorful markings of the gleaming shell beneath.

▼ **Body plan of a gastropod,** here a lung-breathing freshwater snail. Gastropods have adapted to life on land and also to freshwater habitats. Characteristic of gastropods are the flat, creeping foot sole, distinct head with tentacles, mantle, coiled shell made of one piece (univalve), and well-developed alimentary, reproductive, circulatory, nervous and excretory systems.

In most gastropods, rotation (torsion) of the body in the developing embryo (1–3) has brought the opening of the mantle cavity, anus and other organs to the front, and the nervous system is twisted. Sea slugs and bubble shells lose torsion in adult life, and the nervous system of land gastropods is not twisted. It has been suggested that torsion provides the larva with a space (the mantle cavity) into which it can quickly contract its head. This may enable it to sink out of danger quickly or help it to settle on the seabed.

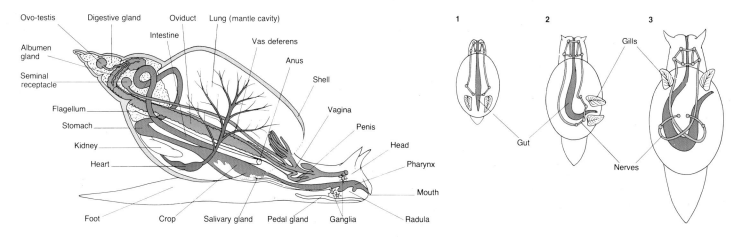

of the primitive ancestor than the more highly evolved gastropods. The mantle cavity and anus are situated toward the rear. The mantle extends forward and houses several pairs of gills, rather more than in most other classes of mollusks. Chitons resemble monoplacophorans and the more primitive groups of prosobranchs, such as the ormers and the top shells, in the ladder-like nervous system with paired ventral nerves and cross-connections. Sexes are separate in the chitons and the eggs are fertilized externally.

Slugs, snails and whelks (gastropods)

This largest class of mollusks contains three-quarters of the living species and shows the greatest variation in body and shell form, function and way of life. Unlike the other classes, which are all aquatic, gastropods have also adapted very successfully to life on land and they have achieved greater diversity in freshwater than the bivalves. Gastropods occur in all climatic zones of the world, colonizing the sea, brackish and freshwater and land. They are among the earliest molluskan fossils.

In the more primitive prosobranchs, gastropods with a cap-like shell, such as slit shells and ormers, which have a trochophore larva with only a short pelagic phase, torsion occurs after the larva has settled. In the more advanced prosobranchs, such as winkles and whelks, the newly hatched veliger larva already has the mantle cavity to the front. Sea slugs and bubble shells do not retain torsion in adult life; loss of shell was influential in the development of this trend. Land and pond snails and slugs have retained torsion, but their nervous system is not twisted.

For the larva, torsion may provide a space into which the animal can quickly contract, enabling it to sink out of danger, or to reach the bottom for settling when it metamorphoses. Advantages of torsion for the adult may include the use of mantle cavity sensors for testing the water ahead or possibly the intake of cleaner water not stirred up by the foot, and providing a space into which the head can be withdrawn. Besides the looping of the alimentary canal and reorientation of the reproductive organs, torsion causes the twisting of the prosobranch nervous system (streptoneury).

The spiral coiling of the gastropod shell is a separate phenomenon from torsion. Coiling occurs also in some of the cephalopods (eg nautiluses), which do not exhibit torsion, and is a way of making a

shell compact. If the primitive cap-like shell had become tall it would have been unstable as the animal moved along. Spiral coiling is brought about by different rates of growth on the two sides of the body. During their evolutionary history, various spirally coiled mollusks in the gastropods, extinct ammonites and nautiloids have uncoiled, producing loosely coiled shells, tubular forms or, in for example limpets, a return to the cap-like shape.

The **prosobranchs** (subclass Prosobranchia) include most of the gastropod seashells—limpets, top shells, winkles, cowries, cones and whelks—as well as a number of land and freshwater species. The prosobranchs, or operculates, have separate sexes, unlike the two other subclasses of gastropod which are hermaphrodite, and the mouth of the

▷ **A dorid sea slug glides** OVERLEAF through a zoological garden. Some members of the family Doridae, called sea lemons, can defend themselves against predators by emitting acid from glands in the mantle.

◁ **Hermaphrodites' courtship embrace.** These two Roman snails will fertilize one another's eggs with an exchange of sperm packets or spermatophores.

▼ **Aeolid sea slug** (*Hermissenda* species) in a Californian tidepool. Members of the family Aeolidiidae use spare parts from their prey, which include jellyfishes and sea anemones. The tentacles of such cnidarians carry nematocysts, stinging structures responsible for the venomous sting of, among others, the Portuguese man-of-war. When a sea slug eats a cnidarian, the victim discharges up to half its nematocysts, but the remainder may reach the papillae secondary organs of breathing on the sea slug's back, via sacs opening off the digestive gland. Later the "live" nematocysts reach the outer skin layer, where the sea slug uses them "second hand" to defend itself against predators.

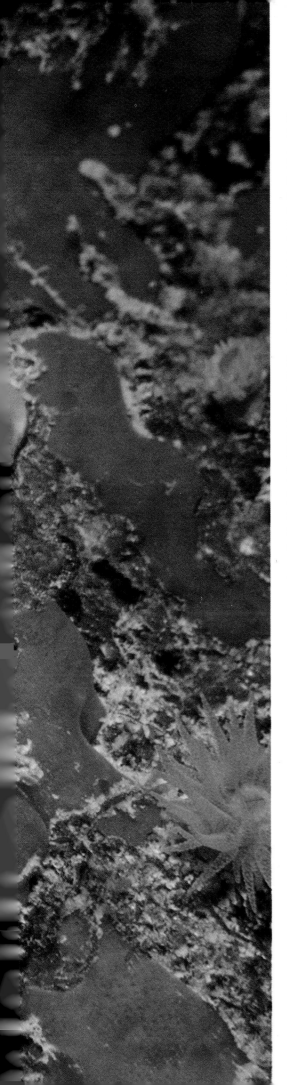

shell is usually, except in limpet forms, protected by a lid or operculum (see below). Aquatic prosobranchs have a gill in the mantle cavity, together sometimes, especially in the carnivorous groups, with a chemical sense organ (osphradium) and slime-secreting hypobranchial gland. The mantle and associated structures exhibit torsion and as a result of this the principal nerves are twisted into a figure-8 in the more highly evolved orders (Mesogastropoda and Neogastropoda).

In marine species there is usually a planktonic trochophore or veliger larval stage which helps to distribute the species.

Prosobranchs live in sea, brackish and fresh water and on land and have an ancient fossil history going back to the Cambrian. Terrestrial prosobranchs are more abundant and varied in the tropical regions than in temperate zones. Shells vary in shape from the typical coiled snail shell to the cap-like shells of limpets and the tubular form of the warm-water vermetids (family Vermetidae) that look from the shell more like marine worm-tubes than mollusks. This diverse group exploits most opportunities in feeding from a diet of algal slime, seaweed, detritus, suspended matter and plankton (ciliary feeders), to terrestrial plants, dead animal matter and other living animals.

The lid or operculum is secreted by glands on the upperside of the back of the foot. It is the last part of the animal to be withdrawn and therefore acts as a protective trapdoor. It keeps out predators and also prevents water loss in land prosobranchs and intertidal species. The operculum is present in the veliger larvae, even in limpets and slipper limpets which later lose the operculum.

The opercula of most prosobranchs are horny, but hard calcareous ones are found in some of the turban and pheasant shells. The thick operculum of the Tapestry turban shell is green and can be about 1in (2.5cm) across: this is the "cat's eye" used for jewelry. The much smaller pheasant shell *Tricolia pullus* from northern Europe and the Australian pheasant shell and others have conspicuous white calcareous opercula.

The different shapes of opercula usually fit the form of the mouth of the shell. Shells with a narrow aperture like cones, for example, have a tall narrow operculum. In some species, such as the whelk *Bullia tahitensis*, there are teeth on the operculum. In the Pink conch shell these teeth are thought to be defenses against attack by predators that include tulip shells. In some species, particularly the land prosobranchs, the operculum may indeed be small, enabling the animal to retreat further inside its shell.

Sea slugs and bubble shells (opisthobranchs) are hermaphrodite—both male and female reproductive systems function in the same individual—and usually have a reduced shell, or none at all. Bubble shells do have a normal external shell, which in the Acteon shell looks very like that of a prosobranch, as it is fairly solid with a distinct spire. Most of the bubble shells, such as *Hydatina*, *Bullaria* and canoe shell species, have an inflated shell which consists mostly of body whorl with little spire, is rather brittle and houses a large animal. Other opisthobranchs have a reduced shell that is internal, for example the thin bubble shells of *Philine* and *Retusa* species. The sea hares have a simple internal shell plate in the mantle that is largely horny. A few species have a bivalve shell (see below). The rest of the sea slugs have lost their shell altogether. They include *Hedylopsis* species with hard spicules in the skin, *Hermaea* species and other sacoglossans such as *Limapontia*, and the large group of the nudibranchs (meaning "exposed-gills") or sea slugs, including the sea lemons, the family Aeolidiidae and Dendronotidae, and many others.

The bodies of opisthobranchs, particularly nudibranchs, can be very colorful. Although they may function as warning coloration or camouflage, little is known of the function of such bright colors, which are less vivid at depth under water. Some sea lemons emit acid from glands in the mantle as a defense against predators.

Opisthobranchs reproduce by laying eggs, often in conspicuous egg masses. The eggs hatch to veliger larvae.

Some species of bubble shells, such as *Retusa* and Acteon shells, have a blunt foot which they use to plow through surface layers of mud or sand. The round shell-less sea lemons creep slowly on the bottom with the flat foot, but many of the sea slugs are agile and beautiful swimmers, capable of speed. Sea butterflies or pteropods swim in surface waters of the oceans.

About 25 years ago a malacological surprise came to light—an animal with a typical bivalve shell but a gastropod body, complete with flat creeping sole and tentacles. This was *Berthelinia limax*, found living on the seaweed *Caulerpa* in Japan. Other bivalve gastropods have since been discovered, also on seaweed, in the Indo-Pacific and Caribbean as well as Japanese waters. They are classed with the sea slugs of the order Ascoglossa.

Bivalve gastropods have a single-coiled

shell in the veliger larva. In mature shells this is sometimes retained at the prominent point (umbone) of the left-hand valve. This development of a bivalve shell in gastropods is an example of convergent evolution rather than evidence of an ancestor shared with the class Bivalvia.

In **pond snails, land snails and slugs** (subclass Pulmonata) the mantle wall is well supplied with blood vessels and acts as a lung. In parallel with this specialization, the lung-breathing snails have specialized in colonizing land and freshwater, although a few continue to live in marine habitats. Like the sea slugs and bubble shells, pulmonates are hermaphrodite, with a complex reproductive system: the free-swimming larval stage is lost in land and freshwater species and, except in the marine genus *Salinator*, there is no operculum.

The shell is usually coiled, although the varied shapes include the limpet form and in several unrelated families the shell is reduced or lost altogether, leading to the highly successful design of slugs. The thin shell of land snails still offers protection against drying out but is more portable and demands less calcium than do the shells of marine gastropods. Both snails and slugs further conserve body water by being active chiefly at night and by their tendency to seek out crevices. The shell-less slugs are freed from restriction to calcareous soils and can also retreat into deeper crevices.

The body is differentiated into a head with tentacles (one or two pairs), foot and visceral mass. The mantle and mantle cavity are at the front (still showing signs of torsion) but the entrance is sealed off except at the breathing pore (pneumostome), which can open and close. The mouthparts and their muscles (buccal mass) may incorporate both a radula and a jaw. Pulmonates are predominantly plant feeders although there are a few carnivores.

The subclass may be divided into three superorders. In the mostly tropical Systellommatophora the mantle envelops the body. The pond snails (superorder Basommatophora) have eyes at the base of their two tentacles, while the land snails and slugs have two pairs of tentacles and eyes at the tips of the hind pair.

Pulmonates succeed in less stable environments than the sea by their opportunistic behavior and the fact that they can enter a dormant state during adverse periods of cold (hibernation) or drought (aestivation). A solidified plug of hardened mucus (epiphragm) can seal off the mouth of the shell and in some species, such as the Roman snail, becomes hardened. Unlike the operculum of a prosobranch, the epiphragm is neither permanent nor attached by muscle tissue to the animal.

Tusk shells

The tusk shells are a small group (Scaphopoda) of around 350 species which are entirely marine and live buried in sand or mud of fairly deep waters. Only their empty shells are to be found on the beach. Tusk shells or scaphopods occur in temperate as well as tropical waters: the large Elephant tusk shell can be up to 4–5in (10–13cm) long. There are two families, the Dentaliidae, which include the large examples more commonly found, and the Siphonodentaliidae (eg *Cadulus* species), which are shorter, smaller and less tubular in shape. The oldest fossil tusk shells known from the Ordovician period 500–440 million years ago. Like the chitons, they have changed little and show very little diversity of body form and way of life.

The shell is tubular, tapering, curved and open at either end. Scaphopods position themselves in the sand with the narrow end protruding above the surface, and through this pass the inhalant and exhalant currents of seawater, usually in bursts rather than as a continuous flow. The broader end of the shell is buried in the sand. From it the head and foot emerge: the foot creates a space in front into which the animal extends the tentacles of the head that pick up detritus, foraminiferans and other microorganisms from the sand. The tentacles are sensory as well as collecting food and conveying it to the mouth. Food can be broken up by the radula, and the shells of foraminiferans are further crushed by plates in the gizzard.

▶ **Wedged in rocks and corals** of shallow Red Sea waters, this giant clam (*Tridacna* species) filter feeds like other bivalves but also houses algae in its brightly colored mantle edge. The algae photosynthesize sugars and starches that are consumed by the clam, which in turn provides shelter and mineral nutrients for its symbiotic partners.

▲ **Mediterranean Fan mussel** or Pen shell (*Pinna nobilis*) is anchored to gravel by strong byssus threads secreted by a gland in the foot. This golden thread was once used in the manufacture of Cloth of Gold, items of which are still to be seen in museums.

▼ **Groups of sensory tentacles** and tiny eyes alternate along the margin of bivalves such as this scallop (*Pecten* species) in the Galapagos Islands. The mantle margin takes over many of the functions fulfilled by the developed head present in other mollusks but lacking in bivalves.

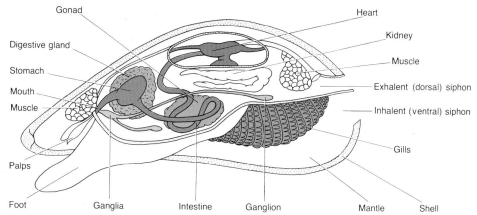

Body plan of a bivalve:

Gonad · Heart · Kidney · Muscle · Digestive gland · Stomach · Exhalent (dorsal) siphon · Mouth · Inhalent (ventral) siphon · Muscle · Gills · Palps · Foot · Ganglia · Intestine · Ganglion · Mantle · Shell

▲ **Body plan of a bivalve.** The bivalve body consists of the mantle, often extended into one or two siphons, the visceral mass and relatively small foot. It lacks a developed head. Usually the siphons and foot can be seen protruding from the shell, but in mussels (*Mytilus*, illustrated above) they largely remain inside.

The anatomy of scaphopods is rather simple. The tube is lined by the mantle. There are no gills, oxygen being taken up by the mantle itself, which may have a few ridges with cells bearing tiny hair-like cilia that help to create a current. Oxygen may also be taken in through the skin of other parts of the body. There are separate sexes, fertilization takes place externally in the sea, and the egg hatches into a pelagic trochophore larva.

Clams, mussels, scallops (bivalves)

Members of this, the second largest class of mollusks with around 15–20,000 living species, are recognized by their shell of two valves which articulate through a hinge plate of teeth and a horny ligament which may be inside or outside the shell.

The bivalve shell can vary considerably in shape, from circular, as in dog cockles, to elongate, as in razor shells. It can be swollen (eg cockles) or flat (eg tellins) and can have radial or concentric shell sculpture (ridges, knobs and spines), bright colors and patterns.

The shell is closed by adductor muscles passing from one valve to the other. Where these attach to the shell, distinctive muscle scars are formed on the inside of the shell. The muscle scars are very important in the

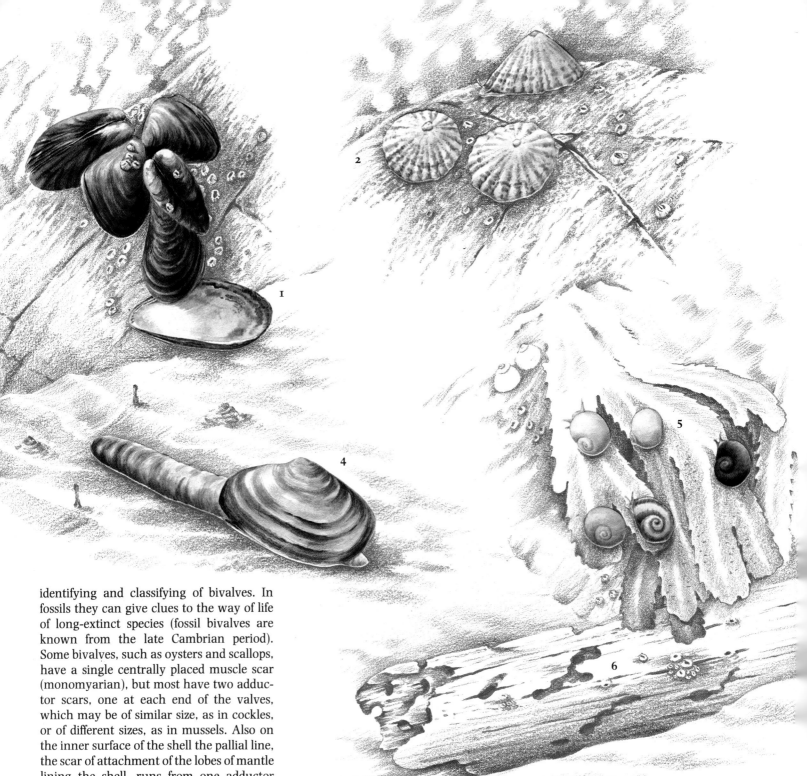

identifying and classifying of bivalves. In
fossils they can give clues to the way of life
of long-extinct species (fossil bivalves are
known from the late Cambrian period).
Some bivalves, such as oysters and scallops,
have a single centrally placed muscle scar
(monomyarian), but most have two adductor scars, one at each end of the valves,
which may be of similar size, as in cockles,
or of different sizes, as in mussels. Also on
the inner surface of the shell the pallial line,
the scar of attachment of the lobes of mantle
lining the shell, runs from one adductor
muscle scar to the other. In those bivalves
with small projecting mantle tubes
(siphons), living on the surface or in shallow
burrows, the pallial line is unbroken and
parallel to the ventral margin of the shell,
opposite the hinge. In the shells of burrowers (eg venus shells and tellins) which
have a long siphon, the pallial line is inden-
ted to provide an extra area for attachment
of the muscles involved in contracting the
siphon when the animal withdraws.

In most bivalves the pair of gills is large
and fills the mantle cavity, performing a
dual role of respiration and feeding. The
primitive nut shells, however, have small
gills which are respiratory only—nut shells
shovel detritus into the mouth with a pair
of labial palps. The carnivorous and more
highly evolved dipper clams (septibranchs,
eg *Cuspidaria* species) have replaced gills
with a wall which controls water flow into
the mantle cavity. They feed on very small
crustaceans and worms drawn in with
water.

The bivalve reproductive system is very
simple. The sexes are usually separate,
although some, like oysters, do alter sex during their lives. The eggs are fertilized
externally, in the sea or in the mantle cavity,
by sperm taken in with surrounding water.
There is a pelagic bivalve veliger larva in
most species.

▲ **Some further representative species of
mollusks.** (1) *Mytilus edulis*, the Edible mussel.
It lives attached to rocks on the lower shore
(3.9in, 10cm). (2) *Patella vulgata*, a limpet;
N Atlantic (2.8in, 7cm). (3) *Janthina exigua*, a
violet sea snail; N Atlantic (0.6in, 1.5cm high).
(4) *Mya arenaria*, a clam or gaper (5.9in,
15cm). (5) *Littorina obtusata*, the flat
periwinkle; NW Europe (0.4in, 1cm). (6) Wood
bored by *Teredo navalis*, the Common ship
worm. (7) *Nautilus pompilus*, the pearly
nautilus (7.9in, 20cm).

▶ **European Common piddock** (*Pholas
dactylus*) in fossilized wood, showing edge of
ridged shell and rounded tip of foot.

3

7

Most species live in the sea, but some have colonized brackish and freshwater. The adults lead a relatively inactive life buried in the substrate, or firmly attached to rock by cement or byssus threads, or boring into stone and wood. A few, like scallops and file shells, flap the valves and swim by jet propulsion.

Octopuses, squids, cuttlefishes, nautiluses
Cephalopods are quite different from the rest of the mollusks in their appearance and their specializations for life as active carnivores. The estimated 650 living species are all marine and include pelagic forms,

swimmers of the open sea, and bottom-dwelling octopuses and cuttlefish. While octopuses can be found in rock crevices on the lower shore, most cephalopods usually occur further out and some penetrate deep abyssal waters, like *Vampyroteuthis infernalis*, which lives 0.3–3mi (0.5–5km) below the surface.

The cephalopods that flourished in the seas of the Mesozoic period over 65 million years ago included nautiloids, ammonites and belemnites. With the exception of nautiluses, these groups, most of which possessed shells, are now extinct.

Most modern cephalopods are descended from the extinct belemnites, which had internal shells.

Cephalopods are typically good swimmers, catching moving fish, and have evolved various buoyancy mechanisms (see p134). They are very responsive to stimuli, due to special giant nerve fibers (axons) with few nerve cell junctions (synapses). This enables messages to pass quickly to and from the brain. (Giant axons are also found

in annelid worms and some other invertebrate groups.) The well-developed cephalopod eye focuses by moving its position rather than changing the shape of the lens. The high metabolic rate of cephalopods is also aided by a particularly efficient blood system with arteries and veins (other mollusks have arteries only) and extra branchial hearts.

All cephalopods except nautiluses have an ink sac opening off the rectum which contains ink, the original artists' sepia. Discharged as a cloud of dark pigment, this confuses an enemy. The cephalopod can also change its color while escaping. Body color and tone are changed by means of pigment cells (chromatophores) in the skin. These are operated under control of the nervous system by muscles radiating from the edge of the chromatophore which can contract it, concentrating the pigment. Stripes and other patterns appear in the skin of cephalopods under certain circumstances.

The possibility that these are a means of communication, for example in recognizing

---

Bivalve Borers

An important number of bivalves, from seven different superfamilies, have adapted from burrowing into soft sand and mud to boring into hard surfaces including mudstone, limestone, sandstone and wood, the wood-boring habit being the most recent to evolve.

Boring developed from bivalves settling in crevices which they subsequently enlarged— one of the giant clams, *Tridacna crocea*, does this. Rock borers of the genus *Hiatella*, although able to use crevices, also erode tunnels of circular section. They push the shell hard against the wall of the burrow by pressure of water in the mantle cavity. At low tide, the red siphons can be seen protruding from holes in the rock low on the shore—they are sometimes known as red-noses.

Most rock borers make their tunnels mechanically by rotation of the shell, often aided by spikes on the shell surface, which erodes away at the rock. The foot may attach the animal to the end of the burrow, as in piddocks. Closure of the siphons helps to keep up fluid pressure. Rock raspings are passed out from the mantle as pseudofeces. While some borers form a tunnel of even width, flask shells (species of *Gastrochaena*) are surrounded by a jacket of cemented shell fragments and live in a rock tunnel that is narrower at the entrance. The contracted siphons dilate outside the shell to form an anchorage during boring.

The date mussels of warm seas and other bivalves have an elongated smooth shell. They make round burrows in limestone by an acid secretion from mantle tissue which is applied to the end of the burrow. The thick,

shiny brown periostracum protects the shell from the mollusk's own acid.

Among the genera of rock borers, *Botula*, *Platyodon* (gapers) and false piddocks drill in clays and mudstone, *Fungiacava* species in coral, and rock borers, flask shells, piddocks and oval piddocks, in rock. The piddocks are recognized by a projecting tooth (apophysis) inside, to which the foot muscles are attached.

Wood piddocks and shipworms burrow into wood. Wood in seawater is merely a transitory habitat, and boring bivalves have therefore adapted in many ways to make the most of what may be a short stay—high population densities, early maturation, prolific breeding and dispersal by pelagic veliger larvae.

Wood piddocks use the wood only for protection, feeding on plankton by normal filtration of seawater, while shipworms exploit the wood further by ingesting the shavings, from which with the aid of cellulase enzymes, they obtain sugar. Shipworms also feed by filtration, but in some species where most food comes from the wood gills are reduced in size.

sex, is being investigated. In bottom-living species like cuttlefish (*Sepia* species), the chromatophores function as camouflage.

The sexes are separate and the male fertilizes the female by placing a sperm package (spermatophore) inside her mantle cavity, where the sperm are released. The yolky eggs are large and there is often a pelagic stage like a miniature adult. Cuttlefish come to inshore waters to breed and after egg laying the spent bodies may be cast up on beaches.

Nautiluses have a brown and white coiled shell. When the nautilus is active, some 34 tentacles protrude from the brown and white coiled shell, and to one side of these are the funnel and hood. The hood forms a protective flap when the nautilus retreats into its shell.

Cuttlefishes are flattened and usually have a spongy internal shell. They often rest on the sea bottom but also may come up and swim. They have 10 arms, eight short and two longer ones, as in squids, for catching prey. The internal shell or "bone" is often found washed up on the seashore.

Squids are torpedo-shaped, active, and adapted for fast swimming. Unlike cuttlefishes, which are solitary, squids move around in shoals in pursuit of fish. The suckers on the 10 arms may be accompanied by hooks. The internal shell or pen is reduced to a thin membranous structure. Oceanic or flying squids can propel themselves through the surface of the water.

Octopuses have adapted to a more sedentary life-style, emerging from rock crevices in pursuit of prey. They can both swim and crawl, using the eight arms. Female octopuses often brood their eggs. The female Paper nautilus is rather unusual in secreting, from two modified arms, a large, thin shell-like egg-case in which she sits protecting the eggs. The male of the species is very small, only one-tenth of the size of the female, and does not produce a shell.

Learning in the Octopus

Octopuses have memory and are capable of learning. A food reward, coupled with a punishment of a mild electric shock, has successfully been used to train octopuses to respond to sight and touch. Sight is important to cephalopods in the recognition of prey. The well-developed eye of cephalopods approaches the acuity of the vertebrate eye more closely than that of any other invertebrate. Presented with two distinct shapes, one leading to a food reward and the other to an electric shock, the octopuses in the above test learned the "right" one after 20–30 trials, although they found some shapes easier to recognize than others, and mirror images of the same shape difficult to separate.

Similar experiments using only the tactile stimulus of cylinders with different patterns of grooves, have shown that octopuses distinguish between and respond differently to rough and smooth objects, different degrees of roughness, and objects with differing proportions of rough and smooth surfaces. Touch is perceived by tactile receptors in the octopus's suckers. Octopuses do not distinguish between objects of different weight, but they can recognize sharp edges and distinguish between inanimate objects and food.

Other researches into the brain and nervous systems of octopuses have led to important advances in knowledge of how nervous systems work in general.

▶ **Prey's-eye view** of a Pacific octopus (*Octopus dofleini*) off the Oregon coast. A red flush replaces the usual yellowish-gray to mottled brown when the octopus is excited.

▼ **Body plan of a cephalopod** (cuttlefish) typically includes mantle, mantle cavity with funnel opening, gills, radula, sharp parrot-like beak, suckered arms (eight short arms, two longer tentacles in cuttlefishes and squid BELOW) and a rounded or, in fast swimmers, pointed torpedo-shaped body. There is usually a reduced internal quilt-like shell or none. A prominent head region is marked by mouth, eyes, arms and cartilage-protected brain.

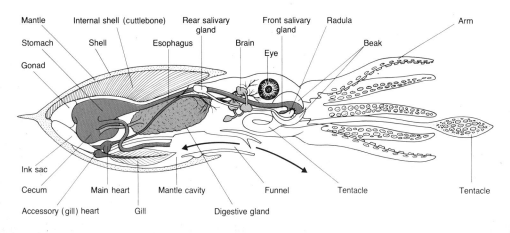

Mantle | Internal shell (cuttlebone) | Rear salivary gland | Front salivary gland | Radula | Arm

Stomach | Shell | Esophagus | Brain | Beak

Gonad | Eye

Ink sac

Cecum | Main heart | Mantle cavity | Funnel | Tentacle | Tentacle

Accessory (gill) heart | Gill | Digestive gland

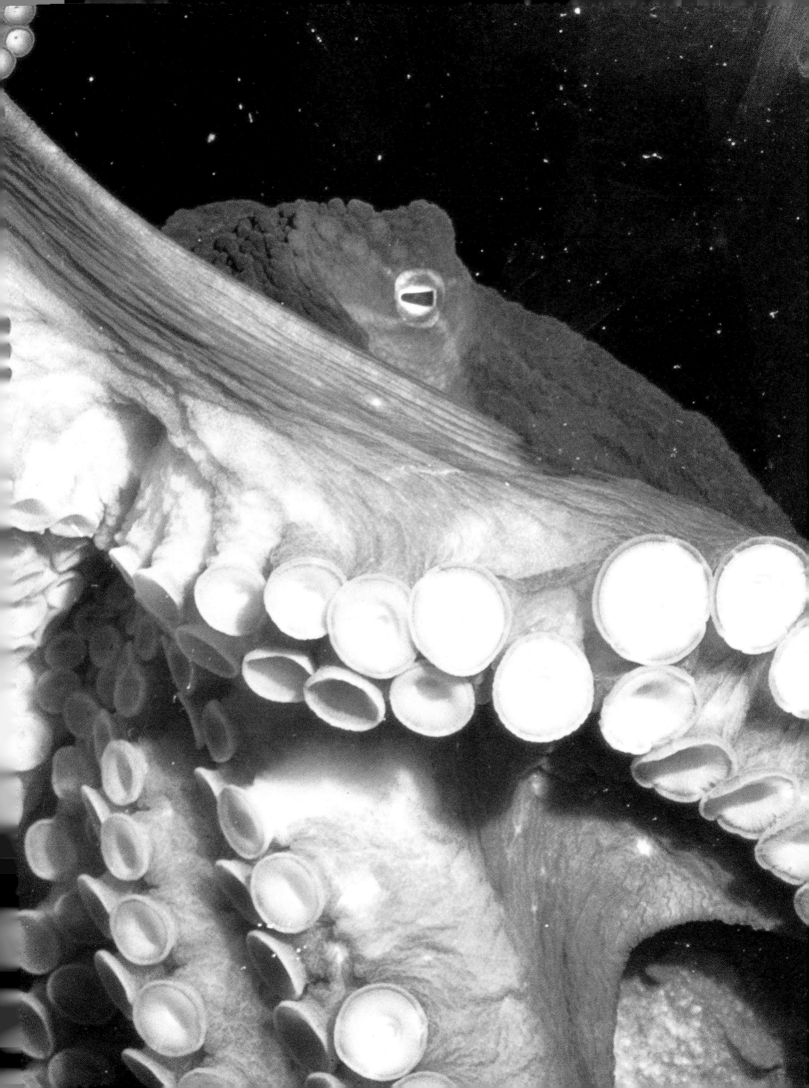

Ecology of mollusks

Mollusks have colonized the sea, fresh water and land. Tropical regions tend to have a more diverse fauna than temperate belts, although temperate New Zealand has one of the richest molluskan land faunas.

Marine habitats can include rocky, coral, sandy, muddy, boulder and shingle shores and also the transitions between freshwater, sea and land found in salt marshes, brackish lagoons, mangrove swamps and estuaries. Beyond the molluskan fauna of the shores is that of the ocean, with communities below the lowtide mark in comparatively shallow water of continental shelves. There is a mosaic of different types of communities on the sea floor, mostly relating to differences in bottom materials. Certain mollusks, like squids, can form part of the free-swimming animal population (nekton) in open water. The veliger larvae of most marine mollusks float passively in the upper waters of the sea, part of the plankton. A few prosobranchs (violet sea snails and heteropods) and opisthobranchs (pteropods or sea butterflies) spend their entire adult lives on or just below the surface as part of the pelagic community. Deep waters of the abyss were once thought to be devoid of life but investigations have revealed a limited but characteristic fauna. With some exceptions (eg squids) abyssal mollusks are small.

Freshwater habitats colonized by mollusks include running waters of streams and rivers, still waters of lakes, ponds, canals, and temporary waters of swamps, all with their own range of species. There are both bivalves and gastropods living in freshwater, the latter including both prosobranchs and aquatic pulmonates. Foreign species can spread dramatically like the freshwater Fingernail clam *Corbicula manilensis*, introduced from Asia to the USA, where it now clogs canals, pipes and pumps. Only the gastropods have successfully colonized land; they include both prosobranchs and pulmonates, although the latter, as both slugs and snails, are the most common in temperate climates.

In the food chains of the sea, mollusks are eaten by other mollusks, as well as by starfish and by bottom-living fish such as rays. Some starfish are notorious predators of commercial mollusks such as oysters and mussels. Some whales eat large quantities of squid. On the shore seabirds probe in mud for mollusks which can form a substantial part of their diet. Such predation by animals which are part of the natural ecosystem can usually be tolerated by mollusks, as they can be prolific breeders.

A few mollusk species have adopted the parasitic way of life. They are nearly all gastropods, with a few bivalves. Among the prosobranchs are parasites on the exterior of the host (ectoparasites), such as needle whelks (eulimids), which are parasites of echinoderms but look like normal gastropods. Internal parasites (endoparasites) are less active and have reduced body organs and less shell. The cap-like genus *Thyca* lives attached to the underside of starfish, in the radial groove under the arms. *Stilifer*, which penetrates skin of echinoderms, has a shell but it is enclosed in fleshy flaps of proboscis (pseudopallium) outside the skin of the host. The further inside the host a parasite is, the more the typical molluskan structure is lost.

Empty gastropod shells are regularly used by hermit crabs (see p100). Commensalism, in which one partner feeds on the food scraps of the other, is shown by the small bivalves *Devonia perrieri*, *Mysella bidentata* and *Montacuta* species which live with sea cucumbers, brittle stars and heart urchins respectively. Symbiosis is demonstrated by the presence of algae (zooxanthellae) in the tissues of sea slugs such as *Elysia* and *Tridachia* species and also in the mantle edge of the Giant clam.

Mollusks are also hosts to their own parasites, many of which may have become established via commensalism. Most parasites of mollusks are larvae of two-suckered flukes. Two commercially and medically important parasites are the liver fluke of sheep and cattle (*Fasciola hepatica*) and blood

▲ **Various oyster species are cultivated** for food in Australia (ABOVE), New Zealand, Japan, the USA and Europe. Japanese pearl oyster culture is also particularly well known. Many natural oysterbeds are all but fished out.

▶ **Mollusk eats echinoderm** in this attack by a Mediterranean Triton shell *Charonia nodifera*. In the food chains of the sea, starfishes are in turn noted predators of bivalves such as mussels and scallops.

▼ **Bird-like flapping** of lateral fins is the characteristic swimming stroke of the Atlantic and Mediterranean cuttlefishes (*Sepiola* species). Much smaller than the torpedo-shaped *Loligo* squids, they may escape predators either by burrowing into the substrate, using the water jet from the mantle funnel, or else by emitting a distracting puff of ink to cover a retreat. When fleeing, the squid abruptly becomes nearly transparent.

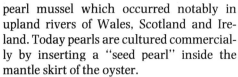

flukes or bilharzia of humans (see p64). Mollusks can also be parasitized by arthropods. Familiar examples are the small white mites *Riccardoella limacum* found crawling on the skin of slugs and snails and the small pea crab, *Pinnotheres pisum*, living in the mantle cavity of the Edible mussel.

Mollusks have long been used in human culture for food, fishing bait and hooks, currency, dyes, pearl, lime, tools, jewelry and ornament. Mother-of-pearl buttons were once manufactured from the shells of freshwater mussels, particularly in the USA, where these mussels were originally common in the rivers. Most pearls come from marine pearl oysters, but at one time fine pearls were obtained from the Freshwater

pearl mussel which occurred notably in upland rivers of Wales, Scotland and Ireland. Today pearls are cultured commercially by inserting a "seed pearl" inside the mantle skirt of the oyster.

Mediterranean cooks are famous for their seafood dishes which utilize gastropods such as necklace shells, top shells, ormers, murexes and occasionally limpets; bivalves such as mussels, scallops, date mussels, venus shells, carpet shells, wedge shells, and razor shells; and also cephalopods including cuttlefishes, squids and octopuses. The traditional "escargot" of French cuisine is the pulmonate Roman snail. On the east coast of the USA the hard-shell clams used by cooks are quahogs (introduced from transatlantic liners to the Solent in southern England), while soft-shell clams are from a range of genera including gapers, carpet shells and trough shells.

Both slugs and snails can be pests of agriculture and horticulture. The mollusks are controlled by biological, cultural and chemical methods, the latter being the ones most usually employed.

A few marine mollusks also compete with man's activities. Bivalves like the shipworm bore into marine timbers, and gastropods, including the slipper limpet and oyster drill, are pests of oyster beds.

The major threats to mollusks are destruction of habitats and pollution—the latter being more important for aquatic species, which are subject to crude oil spills, heavy metals, detergents, fertilizer in water run-off from the land, and acid rain from distant industry. Native land snails of deciduous woodland in the American Midwest, for example, are often not able to cope with the more rigorous conditions of cleared land. In consequence, much North American farmland has been colonized by European mollusks, especially slugs, introduced with plants.

In addition to the harvesting of natural populations and measures limiting trading in shells and marine curios, others aim to prevent introductions of non-native species. The International Union for the Conservation of Nature is also producing Red Data books, assembling information on endangered species that can be used in their conservation.

Little is known of the molluskan fauna of some of the potentially richest and most threatened habitats, many of which are delicate and intolerant of disturbance. Hundreds of snail species are likely to be exterminated before they are even described and studied.   JEC

SPINY-SKINNED INVERTEBRATES

Phylum: Echinodermata

About 6,000 species in 4 subphyla (Homalozoa extinct) and 17 classes (12 extinct).
Distribution: worldwide, exclusively marine.
Fossil record: some 13,000 species known, extensively from pre-Cambrian to recent times.
Size: 0.2in–3.3ft (5mm–1m).

Features: adults coelomate (body originally made up of 3 layers of cells), mostly radially symmetrical and 5-sided; no head, body star-shaped or more or less globular, or cucumber-shaped; calcareous endoskeleton gives flexible or rigid support, often extends into spines externally; tentacled tube-feet, associated with water vascular system unique to phylum, used in respiration, food-gathering and usually locomotion (not crinoids); no complex sense organs; nervous tissue dispersed through body; no nephridia; nearly all species sedentary bottom-dwellers; generally separate sexes, fertilization external; fertilized eggs and larvae generally planktonic; larval stages bilaterally symmetrical.

Sea lilies and **feather stars**
Subphylum: Crinozoa

Starfishes, brittle stars and **basket stars**
Subphylum: Asterozoa

Sea urchins and **sea cucumbers**
Subphylum: Echinozoa

▶ **Closing the gap.** These starfishes (*Archaster typicus*) are not actually mating, as copulation is unknown in echinoderms. By coming close together, the chance of the male's sperm fertilizing the female's eggs in the sea is increased. This behavior is not seen in other echinoderms.

▶ **Brittle and spiny** BELOW, the arms of the brittle star are made up of many ossicles which fit together rather like the vertebrae of the chordate spine. This enables them to be flexible but also makes them liable to fracture at the joints.

▼ **Arms spread, a feather star** (*Tropiometra* species) extends its arms to strain suspended food material from the sea.

THE echinoderms are distinct from all other animal types and easily recognizable. The name echinoderm means spiny-skinned, for most members of the group have defensive spines on the outside of their bodies. They are found only in the sea, never having evolved to cope with the problem of salt balance that life in freshwater would impose on them. As adults they virtually all dwell on the seabed, either, like sea lilies, being attached to it, or, like the starfishes, brittle stars, sea urchins and sea cucumbers, creeping slowly over it. These five groups or classes represent the types of echinoderms found living in the seas and oceans today.

For animals relatively high on the evolutionary scale, it is remarkable that a head has never been developed. Echinoderms show a peculiar body symmetry known as pentamerism. This is effectively a form of radial symmetry with the body arranged around the axis of the mouth. Superimposed on this radial pattern is a five-sided arrangement of the body which is well shown in the starfish. The result is that the echinoderm body generally has five points of symmetry arranged around the axis of the mouth. These points are very often associated with the locomotory organs or tube-feet (see p139).

While five-pointed symmetry or pentamerism is largely displayed by most present-day adult echinoderms, it is interesting to note that their larvae are bilaterally symmetrical (ie symmetrical on either side of a line along the length of the animal), and that their primitive ancestors, which appeared in the pre-Cambrian seas, were also bi-laterally symmetrical. The causes of pentamerism are unclear, but some authorities have suggested that it leads to a stronger skeletal framework.

The body of echinoderms shows a deuterostome coelomate level of organization (see p16). This means that they are relatively highly evolved invertebrates with a body constructed originally from three layers of cells.

The echinoderm skeleton is made of many crystals of calcite (calcium carbonate). It is unusual because these crystals are perforated by many spaces in life (reticulate) so that the tissue which forms the crystals actually invades them. Such a reticulate arrangement leads to a lightening of the crystal structure and hence a reduction in weight of the animal without any loss of strength (see right). One side effect of this crystal structure is that it is easily invaded by mineral after the death of the animal and thus it fossilizes beautifully.

The skeleton supports the body wall or test. This reinforced structure may be soft (as in sea cucumbers) or hard (sea urchins) but it should never be thought of as a shell because it is covered by living tissue.

The exterior of each class of echinoderm appears different, and so too is the way in which the skeleton has been deployed. In the sea cucumbers the calcite crystals are embedded in the body wall and linked by flexible connective tissue in a way that does not occur in the other classes. In the starfish there is sometimes a flexible body wall, but more often the crystals are grouped close together, sometimes being "stitched" together by fibers of connective tissue running through the crystal perforations. Individual crystals may be extensively developed to form spines or marginal plates. The sea

urchins have carried the skeletal process further, for in almost all the skeleton is rigid, being composed of many interlocking crystals. At the same time there is a reduction in the soft tissue of the body wall. The sea urchins have some of the most complex arrangements of muscle and skeleton in the phylum, for example the chewing teeth or "lantern teeth," as Aristotle called them, and the pedicellariae.

In the sea lilies and feather stars and the brittle stars the skeleton is massive and arranged as a series of plates, ossicles and spines with a minimum of soft tissue. In both these classes the major internal organs or viscera are contained in a reduced area, the cup-like body (theca) of the crinoids and the disk-like central body of the brittle stars. Here the skeleton reinforces the body wall, which remains flexible; but in the arms the ossicles become massive, operating with muscles and connective tissue in a way rather reminiscent of the vertebrae of the human backbone. In the arms of both types there is relatively little soft tissue. In the sea lilies the arms branch near their bases into two or more main axes, each bearing lateral branches called pinnacles. The arms of brittle stars branch only in the basket stars.

The drifting echinoderm larvae also have a skeleton, which serves to support their delicate swimming processes.

Another unique feature of echinoderms is their water vascular system. This probably arose in the primitive echinoderms as a respiratory system pointing away from the substrate which could be withdrawn inside the heavily armored test. As the echinoderms became more advanced it was arranged around the mouth, but still held away from the substrate. Branching processes developed, forming a system of tentacles that became useful for suspension-feeding as well as respiration. It is in this state that the water vascular system is seen in present-day sea lilies and feather stars. Their branched tentacles, also called tube-feet (although in the crinoids they have no locomotory role), are arranged in a double row along the upper side of each arm, bounding a food groove, and along the branches of the arms (pinnules). The tube-feet can be extended by hydraulic pressure from within the animal, and much of the water vascular system is internal. They are supplied with fluid from a radial water canal which runs down the center of each arm, just below the food groove, and which sends a branch into each pinnule. The radial water canal of each arm connects with that of its fellows via a circular canal running around the gullet of the animal. Pressure is generated inside the system by the contraction of some of the tube-feet, and also by special muscles in the canal itself which generate local pressure increases to distend the neighboring tube-feet. The water vascular system in crinoids is associated with several other tubular networks, notably the hemal and peri-hemal systems (whose role is less easy to define) and the radial water canal also runs close to the radial nerve cord which controls the tube-feet.

The activities of the tube-feet relate to gas exchange (respiration) and food gathering. The tube-feet are equipped with mucous glands in crinoids and when a small fragment of drifting food collides with one, the fragment sticks to the tube-feet, is bound in mucus and flicked into the food groove by which it passes down to the central mouth. The tube-feet are arranged in double rows alternating with small non-distensible lappets. This arrangement assures their efficient use in feeding.

Crinoids exploit currents of water in the sea. They do not pump water to get their food, but gather it passively. They "fish" for food particles using the tube-feet and select mainly those in the 0.01–0.02in (0.3–0.5mm) size-range.

In all the remaining groups of echinoderms the orientation of the body is reversed with respect to the substrate. The tube-feet actually make contact with the ground over which the animals are moving and thus take up an additional role in locomotion. This happens in the starfishes, sea urchins and sea cucumbers but not in the basket and brittle stars, where movement is achieved by bending the arms, while the tube-feet are still important in respiration and food gathering. In the basket stars they are well developed for suspension feeding in a way which has interesting parallels with the crinoids. The basket stars, too, exploit currents of water, and arrange their complex branching arms with tube-feet as a parabolic net sieving the water currents for particles in the 10–30μm size range. Thus they do not exactly compete with the crinoids in the same habitat. They are able to withstand stronger currents than the crinoids. In the remaining types of brittle stars there is a range of feeding habits. Some, like *Ophiothrix fragilis*, are suspension feeders, often living in huge beds. Others are detritus or carrion feeders. In many species the tube-feet, which are suckerless, are very important in transferring food to the mouth and have a sticky mucous coating which

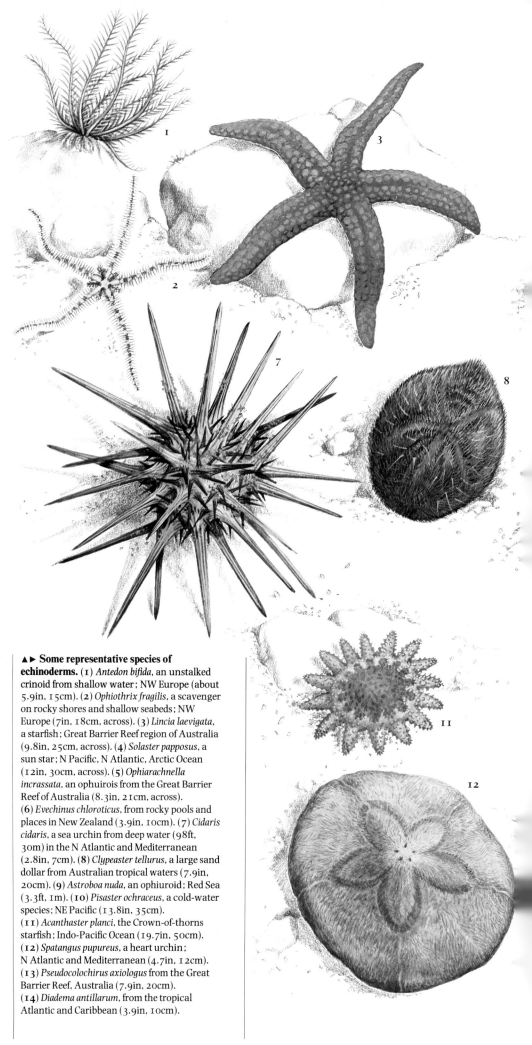

▲ ▶ **Some representative species of echinoderms.** (**1**) *Antedon bifida*, an unstalked crinoid from shallow water; NW Europe (about 5.9in, 15cm). (**2**) *Ophiothrix fragilis*, a scavenger on rocky shores and shallow seabeds; NW Europe (7in, 18cm, across). (**3**) *Lincia laevigata*, a starfish; Great Barrier Reef region of Australia (9.8in, 25cm, across). (**4**) *Solaster papposus*, a sun star; N Pacific, N Atlantic, Arctic Ocean (12in, 30cm, across). (**5**) *Ophiarachnella incrassata*, an ophuirois from the Great Barrier Reef of Australia (8.3in, 21cm, across). (**6**) *Evechinus chloroticus*, from rocky pools and places in New Zealand (3.9in, 10cm). (**7**) *Cidaris cidaris*, a sea urchin from deep water (98ft, 30m) in the N Atlantic and Mediterranean (2.8in, 7cm). (**8**) *Clypeaster tellurus*, a large sand dollar from Australian tropical waters (7.9in, 20cm). (**9**) *Astroboa nuda*, an ophiuroid; Red Sea (3.3ft, 1m). (**10**) *Pisaster ochraceus*, a cold-water species; NE Pacific (13.8in, 35cm). (**11**) *Acanthaster planci*, the Crown-of-thorns starfish; Indo-Pacific Ocean (19.7in, 50cm). (**12**) *Spatangus pupureus*, a heart urchin; N Atlantic and Mediterranean (4.7in, 12cm). (**13**) *Pseudocolochirus axiologus* from the Great Barrier Reef, Australia (7.9in, 20cm). (**14**) *Diadema antillarum*, from the tropical Atlantic and Caribbean (3.9in, 10cm).

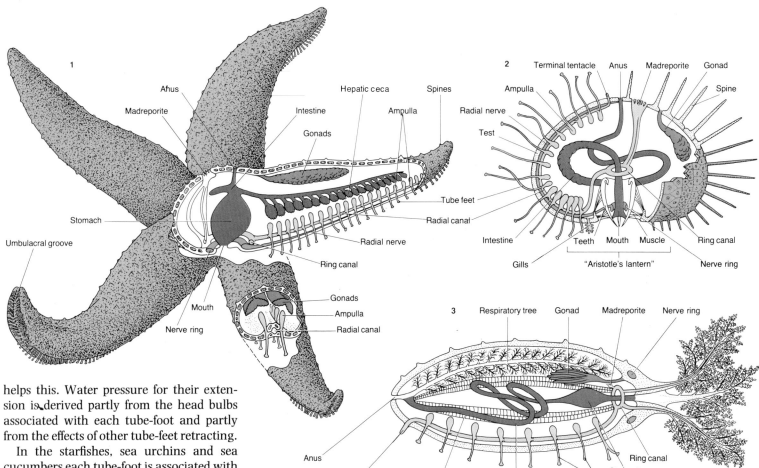

1

Anus
Madreporite
Hepatic ceca
Intestine
Spines
Ampulla
Gonads
Stomach
Umbulacral groove
Tube feet
Radial canal
Radial nerve
Ring canal
Mouth
Gonads
Ampulla
Nerve ring
Radial canal

2

Terminal tentacle   Anus   Madreporite   Gonad
Ampulla   Spine
Radial nerve
Test
Intestine
Gills
Teeth   Mouth   Muscle
"Aristotle's lantern"
Ring canal
Nerve ring

3

Respiratory tree   Gonad   Madreporite   Nerve ring
Anus
Terminal tentacle   Muscle bands   Intestine   Tube feet
Ring canal
Stomach
Tentacles

helps this. Water pressure for their extension is derived partly from the head bulbs associated with each tube-foot and partly from the effects of other tube-feet retracting.

In the starfishes, sea urchins and sea cucumbers each tube-foot is associated with its own reservoir or ampulla. The ampulla is thought to play a role in filling the tube-foot with water vascular fluid. It has its own muscle system and connects to the foot by valves to control the flow. However it seems certain that fluid pressure within the water vascular system is also important. The shafts of the tube-feet are equipped with muscles for retraction and for stepping movements.

Suckered tube-feet occur in all sea urchins and many sea cucumbers. Some asteroids, eg *Luidia* and *Astropecten* species, lack suckers on the tube-feet, and most of these burrow in sand. Other starfishes, inhabiting hard substrates, have suckered tube-feet and use them for locomotion and for seizing prey. In the burrowing sea urchins, eg *Echinocardium* species, some of the tube-feet are highly specialized for tunnel-building and for ventilating the burrow. In the sea cucumbers the ambulacral tube-feet may be used for locomotion and respiration, while those surrounding the mouth have become well developed for suspension or deposit feeding and form the characteristic oral tentacles. There are many closely related sea cucumbers which feed in slightly different ways, each having slightly modified oral tentacles so they can exploit food deposits of detritus particles of different sizes.

The fluid within the water vascular system is essentially seawater with added cellular and organic material. Water

vascular fluid is responsible for other tasks apart from driving the tube-feet. It transports food and waste material and conveys oxygen and carbon dioxide to and from the tissues of the body. It contains many cells. These are mainly amoeboid coelomocytes which have a role to play in excretion, wound healing, repair and regeneration. No excretory organs have been identified in the echinoderms.

The water vascular system of starfishes, basket and brittle stars, and sea urchins appears to communicate to the exterior of

▲ **Body plans of starfish, sea urchin and sea cucumber.**

▼ **Sucker power:** close-up of the tube-feet and spines of the Crown-of-thorns starfish.

▷ **Social stars.** OVERLEAF Brittle stars often gather on the seabed. Here *Ophiothrix quinquemaculata* is grouped with other invertebrates on the floor of the Mediterranean.

▶ **Smothering to death:** the spiny starfish *Marthasterias glacialis* feeds on a sea squirt.

▼ **Test appendages,** a selection of the microscopic tong- or forceps-like organs found between the spines of all sea urchins and most starfishes.

Grooming Tools of Echinoderms

Between their spines, most starfishes and all sea urchins carry unique small grooming organs like microscopic tongs or forceps. These intriguing organs were once thought to be parasites on the tests of the animals which bear them but are an integral part of the animal. Each consists of two or more jaws supported by skeletal ossicles called valves. Some starfish types are directly attached to the test. Others, like those of sea urchins, are carried on stalks, so that the jaws can reach down to the surface of the test.

The jaws are caused to open or close by muscles attached to nerves from both the base of the epithelium of the test and from special receptors responsive to touch and certain chemicals. Most of these receptors are situated on the inside of the jaw blades, but some lie on the outside.

In an undisturbed animal the epithelial nerves may close down most of the pedicellariae (except some on the sea urchins, see below), so that they are inactive with the jaws closed. If an intruding organism strays onto the test, such as a small crustacean or a barnacle larva seeking a place to settle, the resultant tactile stimuli cause the pedicellariae to gape open and thus expose the special touch receptors. If these are stimulated, the jaws rapidly snap shut, trapping the intruder. The pedicellariae of starfish (**1**, skeletal parts only) and the tridentate (in fact, three-jawed)

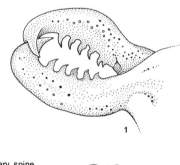

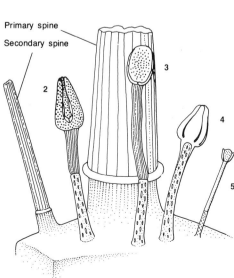

(**2**) and ophiocephalous (snake-head) ones (**3**) of sea urchins are all specialized for such activities and often have fearsome teeth to grip their victims.

In the globiferous (round-headed) pedicellariae (**4**) of the sea urchin class Echinoidea, there are venom sacs. Here the jaws close only on objects which carry certain chemicals. The venom is injected into the victim via a hollow tooth in many echinoids and in species such as *Toxopneustes pileosus* it has a powerful effect. Globiferous pedicellariae detach after the venom is injected and remain embedded in the tissue of the intruder. They seem to be mainly deployed by the sea-urchins as defenses against larger predators such as starfish.

In the sea urchins, sand dollars and heart urchins the smallest class of pedicellariae are known as trifoliate (three-leaved) (**5**). They differ from all the others in having spontaneous jaw movements, "mouthing" over the surface of the test in grooming and cleaning activities. This persists even when these pedicellariae are removed from the test.

Far from being the parasites they were once believed to be, pedicellariae therefore serve to keep the surface of the echinoderm free from other animal, or plant, organisms. Their complex structure is a good example of the intricacies of echinoderm biology, but their various roles are not yet all understood.

the animal via a special sieve-like plate, the madreporite. In sea cucumbers the madreporite is internal, while in the crinoids it is lacking altogether. It used to be thought that seawater entered and left the water vascular system via the madreporite, but more recent research in sea urchins shows that in fact very little water actually moves across this special structure. In starfish and sea urchins the madreporite may be associated with orientation during locomotion (see below).

The nervous system of echinoderms is peculiar to the group. Because of the absence of a head there is no brain and no aggregation of nerve organs in one part of the body. In fact, with the exception of the rudimentary eyes (optic cushions) of starfishes, the balance organs (statocysts) of some sea cucumbers and the chemosensory receptors of pedicellariae of sea urchins (see box (p143) there are no complex sense organs in echinoderms. Instead there are simple receptor cells responding to touch and chemicals in solution. These appear to be widely spread over the surface of the animals. Some authorities even suggest that all the external epithelial cells of starfish and sea urchins may have a sensory function.

In all living echinoderms the main part of the nervous system comprises the nerve cords which run along the axis of each arm close to the radial water canal. These radial nerve cords are linked together around the esophagus by a circum–esophageal nerve cord so that the activities of one arm or ray may be integrated with the activities of the others.

The control of the tube-feet and body-wall muscles is under the command of each radial nerve cord. The responsibilities for coordinated locomotion and direction of movement lie here too. In directional terms echinoderms may move with one arm or ray taking the lead, or even with the space between two acting as a leading edge. Where there is a need to back away, the animal may either go into reverse or actually turn around. In the starfish and sea urchins there is some evidence to suggest that the space between two rays which contains the madreporite may frequently act as the leading edge, possibly because the madreporite has some, as yet unknown, sensory function.

Echinoderms are all very sensitive to gravity and generally show a well-defined righting response if they are turned upside down. It has been suggested that in the sea urchins, small club-like organs, sphaeridia, act as organs of balance. All other echinoderms, apart from a few sea cucumbers, lack the balance organs (statocysts) that are frequently found in mollusks and crustaceans.

In starfishes and sea urchins the outer surface of the body is covered by a well-developed epithelium at the base of which lies a network of nerves. This nerve plexus controls the external appendages of these two groups which are richly developed in many species. The appendages include various effector organs, movable spines, pedicellariae (minute tong-like grasping organs), sphaeridia in sea urchins and paxillae and papulae in starfishes. These organs are concerned with defending the animal against intruders and keeping the delicate skin of the test clean from deposits of silt and detritrus.

The basi-epithelial nerve plexus connects with the radial nerve cords of each arm or ray and forms a system of fine nerves linking the receptor site of the epithelium with the various effector organs.

The various groups of existing echinoderms show characteristic patterns of behavior. All echinoderms are sensitive to touch, and to waterborne chemicals which signal the presence of desirable prey or of potential predators which must be avoided. Most starfish species are efficient predators, feeding on other invertebrates, such as worms, mollusks and other echinoderms. They are able to "smell" the presence of suitable prey in the water and move efficiently towards it.

For the common European starfish, *Asterias rubens*, mussels and oysters are significant prey items. The sun stars *Solaster endeca* and *Crossaster papposus* also feed on bivalves, but may attack *Asterias* too. In tropical waters starfishes show a variety of tastes, but the Crown-of-thorns, *Acanthaster planci*, is well-known for its selection of certain species of reef-building coral as prey. All these echinoderm predators will move efficiently toward the source of chemical scent in the water. When they have arrived at it they will commence attack.

Some of the burrowing starfish (eg *Astropecten* species) will ingest their prey of gastropods whole. *Crossaster* species may attack *Asterias* by hanging on to one ray with the mouth and eating it while the prey drags the predator about. *Acanthaster* feeds on objects too large to be ingested whole, so it everts the stomach membranes through its mouth and smothers the prey with these, digesting the victim outside its body. When the process of digestion is complete the stomach membranes are withdrawn.

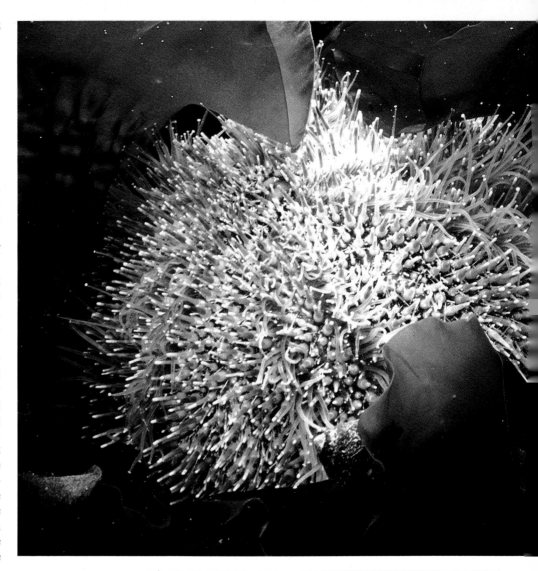

Members of the genus *Asterias*, like other starfishes which prey on bivalves, are able to use their tube-feet with their suckers to prize open the valves of a mussel or oyster. They do this by climbing on to the prey and attaching some tube-feet to each valve. The two valves are then pulled apart by the persistent actions of the tube-feet. The muscles which keep the shells closed eventually tire, so that they gape ever so slightly. A gap of about $\frac{1}{12}$ of an inch is all that is needed for the starfish to insert some of its stomach folds passed out through its mouth. Once this has been done, digestion of the victim will begin and in the end only the cleaned empty shell will remain.

It is interesting to note that in both tropical starfish (eg *Acanthaster* species) and temperate species (eg *Asterias*), solitary individuals display a different type of feeding behavior from that of individuals feeding together in groups. In these starfishes, regular feeding tends to be solitary and at night, the individuals being well spaced one from the next. In some populations, individuals will gather periodically in large numbers at a superabundant food source and feed by day as well as by night; as a result of such social feeding, the growth rate of individuals considerably surpasses the

The 3 Subphyla and 5 Classes of Living Echinoderms

Sea lilies, feather stars
Subphylum: Crinozoa

About 650 species; sedentary, mostly stalked, at least in young even if adults free-living; branching main nervous system; anal opening on same surface as mouth; 6 classes and most species known from fossils. One surviving class, Crinoidea (order Articulata), lacking madreporite, spines and pedicellariae appendages. Includes genus *Antedon*.

Starfishes
Subphylum: Asterozoa

Stemless, mobile, free-living; mouth surface faces down, nervous system on mouth surface; usually have arms (rays).

Class: Asteroidea —**starfishes**

About 2,000 species; flattened, star-shaped with 5 arms (a few with more); endoskeleton flexible; arms contain digestive ceca; branching madreporite.
Subclasses: Somasteroidea (including genus *Platasterias*); and Euasteroidea, 5 orders including Platyasterida (eg genus *Luidia*).

Class: Ophiuroidea —**brittle stars, basket stars**

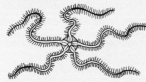

About 1,600 species; flattened, 5 sided with long flexible arms rarely branched, clearly demarked from central "control" disk; madrepore on underside; no anus or intestine; tube-feet lack suckers.
Orders: Oegophiurida (including genus *Ophiocanops*); Phrynophiurida **basket stars** (including genus *Gorgonocephalus*); and Ophiurida **brittle stars** (including genus *Ophiura*).

Sea urchins
Subphylum: Echinozoa

Stemless, mobile, free-living; mouth surface downward or to side; main nervous system on oral surface; without arms or rays.
Six fossil classes, of which 2 are living.

Class: Echinoidea —**sea urchins, sand dollars, heart urchins**

About 750 species; mainly globular or disk-shaped, without arms; covered with numerous spines and pedicellariae; tube-feet usually ending in suckers; endoskeleton comprises close-fitting plates.
Subclass Perischoechinoidea: 1 living order Cidaroida, including genus *Cidaris*.
Subclass Euechinoida: comprises superorders Diadematacea (eg *Asthenosoma, Diadema*); Echinacea (eg *Echinus, Psammechinus, Paracentrotus, Toxopneustes*); Gnathostomata with orders Holectypoida (eg *Echinoneus, Micropetalon*) and Clypeasteroida

(eg *Rotula, Clypeaster, Echinocyamus*); and Atelostomata with suborders Holasteroidea (eg *Pourtalesia, Echinosigria*), and Spatangoida (eg *Spatangus, Echinocardium, Brissopis*).

Class: Holothuroidea —**sea cucumbers**

About 500 species; long, sac-like, without arms; bilaterally symmetrical; mouth surrounded by tentacles; no spines or pedicellariae; endoskeleton reduced to microscopic spicules or plates, or absent.
Subclass Dendrochirotacea: orders Dendrochirotida (eg *Cucumaria, Thyone, Psolus*); and Dactylochirotida (eg *Rhopalodina*).
Subclass Aspidochirotacea: orders Aspidochirotida (eg *Holothuria*); and Elasipodia (*Pelagothuria, Psychropotes*).
Subclass Apodacea: orders Molpadiida (eg *Molpadia, Caudina*); and Apodida (eg *Synapta, Leptosynapta*).

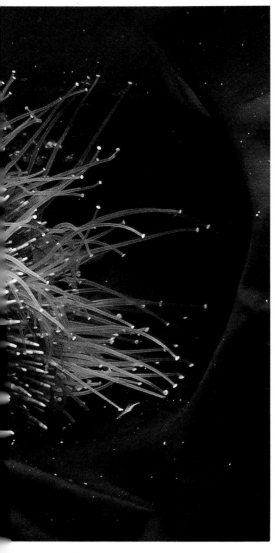

▲ **Sea urchin under pressure.** *Echinus esculentus* was taken for its edible roes and regarded as a delicacy in Tudor England. Now it is fished for its ornamental test.

▼ **Cleaning up.** The sea cucumber *Holothuria forskali* creeps across the seabed using modified oral tube-feet to collect the sediment and the food it contains.

norm. To what extent these differences are acquired or inherited is not yet clear.

In the sea urchins there is a range of feeding behavior. Many of the round (or regular) echinoids (eg the genera *Strongylocentrotus*, *Arbacia* and *Echinus*) are omnivores. They browse on algae and encrusting animals such as hydrozoans and barnacles, using their Aristotle's lantern teeth. In many places echinoids are important at limiting the growth of marine plants and compete very successfully with other types of algal grazers, including gastropod mollusks and fish.

The irregular sea urchins, including the sand dollars and heart urchins, are more specialized feeders. The sand dollars live partly buried in the sand and use their modified spines and tube-feet to collect particles of detritus for food. These are then passed to the mouth along ciliated tracts. The heart urchins burrow quite deeply in sand and gravels and ingest the substrate entire. They have lost the Aristotle's lantern. As the substrate particles pass along their guts any organic material is digested and "clean" substrate is passed out from the anus.

The sea cucumbers have diversified to exploit a number of food sources. In virtually every case the form of their oral tentacles, the specialized tube-feet arranged round the mouth, is adapted to gathering food.

Some groups sweep the surface of sand and mud for particles of detritus and thus live as deposit feeders. Others are suspension feeders relying on currents of seawater to sweep suspended particles of food into their oral tentacles. In both cases the size of the sweeping or filtering fronds and the gap between will dictate the sizes of the particles collected.

The fascinating and often beautiful echinoderms inhabit all the world's seas from the intertidal zone down to the ocean abyss. They are also present in all latitudes from the tropics to the poles. In temperate intertidal zones such as the North Atlantic and North Pacific starfishes and sea urchins are familiar organisms. In some places sea urchins may be harvested for use as food, eg *Paracentrotus lividus* in Ireland and the Mediterranean. Brittle stars, and sea cucumbers, though present, are less conspicuous intertidally. The shallow seas overlying the continental shelves are particularly good habitats for echinoderms where coastal currents, rich in nutrients, sweep detritus and plankton for suspension feeders such as the crinoids and sea cucumbers and nourish prey suitable for starfishes. In tropical areas the development of reefs allows a great diversity of echinoderm species to develop because of the variety of niches. Although all groups of echinoderms inhabit the ocean abyss it is here that the sea cucumbers flourish, often in great densities, moving over the benthic ooze in search of detrital food. Here too some highly unusual epibenthic sea cucumbers have taken to a swimming life, moving along in deep currents and collecting food as they go.

The majority of echinoderm species are dioecious—the sexes are separate. A few are hermaphrodite, passing through a male phase before becoming functional females. In one genus, *Archaster* from the West Pacific, pseudo-copulatory activity occurs, with the one partner climbing on top of the other, but even here fertilization is external. Sperm and eggs are released into the seawater via short gonoducts. In many species this is almost a casual affair, the partners not coming close together. However, synchrony of spawning is essential and this is usually governed by water temperature, and chemical stimuli operating between participants.

Antarctic and abyssal echinoderms often brood their eggs, and a direct development of juveniles occurs in brood pouches or between spines. In the remaining vast majority of species the fertilized eggs drift in the plankton and develop through characteristic larval stages, usually feeding on minute planktonic plants such as diatoms and dinoflagellates. After a period of larval life that ranges from a few days up to several weeks in different species, metamorphosis occurs, and the juvenile echinoderms settle on the sea bed. AC

SEA SQUIRTS AND LANCELETS

Phylum: Chordata

Sea squirts

Subphylum: Urochordata (or Tunicata)
About 2,000 species in 3 classes and 7 families.
Distribution: worldwide; bottom-dwelling and pelagic at all depths and in all oceans.
Fossil record: very few, from the Cambrian to the Quaternary (600–500 million years ago to recent).
Size: individuals from less than 0.4in (1cm) to about 8in (20cm) long.

Features: chordate characteristics—hollow, dorsal nerve cord, enterocoelic body cavity, gill slits, tail behind anus, and notocord—are all present in the larvae; adults show no segmentation, lack hollow dorsal nerve cord and notochord, and most have the gills surrounded by a large cavity (atrium) and are ensheathed in a test or tunic of tunicin, a substance related to cellulose; hermaphrodite; reproduction often involves budding and an asexual phase; life cycles complex in Thaliacea and Larvacea.

Sea squirts

Class: Ascidiacea
Includes genera *Aplidium*, *Botryllus*, *Clavelina*, *Gonia*.

Salps or pelagic turnicates

Class: Thaliacea
Includes genera *Pyrosoma*, *Salpa*.

Class: Larvacea
Sexually mature adults resemble larvae.
Includes genus *Oikopleura*

Lancelets

Subphylum: Cephalocordata (or Acrania)
Fewer than 20 species in 2 families.
Distribution: temperate to tropical shallow sea waters, adults bottom-dwellers, burrowing in sand.
Fossil record: none.
Size: up to 2in (5cm) long.

Features: simple fish-like chordates lacking a recognizable head, with hollow dorsal nerve cord similar to vertebrates, but no vertebrae, instead a notochord, muscle blocks segmented; enterocoelic coelom and well-developed gills; can swim; sexes separate; planktonic tornaria larva produced.

Family: Branchiostomidae, genus *Branchiostoma* (or *Amphioxus*).

Family: Asymmetronidae, genus *Asymmetron*.

▶ **Lowly chordates.** Sea squirts can be single individuals like *Ciona intestinalis* with its translucent body and yellow-fringed siphonal openings, or colonial animals like *Botryllus schlosseri* where a few individuals are grouped in star-like patterns around a common exhalant opening.

CHORDATES are "higher animals" possessing a single, hollow, dorsal nerve cord and a body cavity that is a true coelom. The most familiar chordates are the vertebrates, fishes, amphibia, reptiles and mammals, where a definite backbone and bony braincase are to be found (subphylum Craniata or Vertebrata). However, there are a number of lowly chordates which display the phylum's characteristics at a simple level. All aquatic, they are included in the two other subphyla of the chordates, the Urochordata (sea squirts) and Cephalochordata (lancelets).

The adult **sea squirts** look nothing like vertebrates. They are bottom-dwellers growing attached to rocks or other organisms. Their bodies are encased in a thick tunic made of material which resembles cellulose, the main constituent of plant cell walls. At the top of the body lies the inhalant siphon, and on the side is the exhalant siphon. There is no head. Water is pumped in via the inhalant siphon and passes through the pharynx and out between the gills into the sleeve-like atrium. From the atrium it is discharged via the exhalant siphon. The gills serve as a respiratory surface and also act as a filter, extracting suspended particles of food. They are ciliated, the tiny hair-like cilia providing the pumping force to maintain the respiratory and filter current. Acceptable food particles are collected in sticky mucus secreted by the endostyle, a sort of glandular gutter running down one side of the pharynx, and they are then passed into the gut for digestion. Waste products are liberated from the anus which opens inside the atrium near the exhalant siphon.

Sea squirts may be solitary or colonial. In a colony the exhalant siphons of individuals open into a common cloaca, and each individual retains its own inhalant siphon. Colonial squirts may be arranged in masses, as in *Aplidium* species, where the individuals are hard to recognize, or in encrusting plate-like growth, eg *Botryllus* species, where the individuals can easily be made out.

The heart lies in a loop of the gut and services the very simple blood system. Sea squirts are hermaphrodite, each having male and female organs. In some species the eggs are retained in the atrium, where the sex ducts open, and where they are fertilized by sperm drawn in with the feeding and respiratory currents. The embryos can develop here in a protected environment. In other species the eggs and sperm are both liberated into the sea where fertilization occurs. The embryo quickly develops into a

▶ **Like a row of glass bottles,** these colonial salps drift through the deep waters of the ocean.

▼ **Colonial sea squirts** on the Great Barrier Reef. In some species the green coloring is due to commensal algae growing in their tissues.

tadpole-like larva. This process may take from a few hours to a few days, a time-scale which makes these animals ideal for observation and experimental embryology.

The "tadpoles" of sea squirts are small, independent animals like miniature frog's tadpoles. They are sensitive to light and gravity, enabling the animal to select an appropriate substrate for settlement, attach-ment and metamorphosis. These tadpoles obviously serve to distribute the species too. It is the ascidian tadpole with its chordate characters of dorsal hollow nerve cord, stiff-ening notochord supporting the muscular tail, and features of the head which tells us much more of the likely evolutionary posi-tion of the sea squirts than does the adult.

Some sea squirts are of economic import-ance because they act as fouling organisms encrusting the hulls of ships and other marine structures. They are also of consider-able evolutionary and zoological interest.

The other urochordates, thaliaceans (salps and others), and the larva-like larva-ceans, have characters fundamental to the group but have evolved along pelagic and not bottom-dwelling lines. Members of one thaliacean order, the Pyrosomidae, are com-monly found in warmer waters and emit phosphorescent light in response to tactile stimulation. These pelagic animals (eg *Pyrosoma* species) live in colonies, but some other types exist as solitary individuals. They have complex life-histories. *Oikopleura* is a larvacean genus of small animals like the tadpoles of sea squirts. Rhythmic move-ments of the tail draw water through the openings of the gelatinous "house" in which they live and which is secreted by the body.

The **lancelets** are a minor group of chordates containing only two genera and under 20 species. These small, apparently fish-like animals are known as lancelets because of their elongated blade-like form. They show primitive chordate conditions. The notochord, which in sea squirt larvae merely supports the tail, extends into the head (hence the term cephalochordata). There is no cranium as in the craniates or

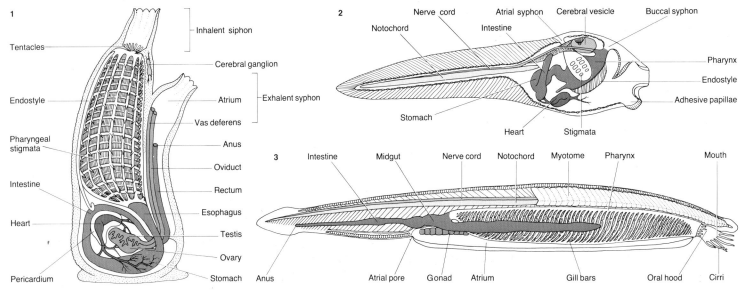

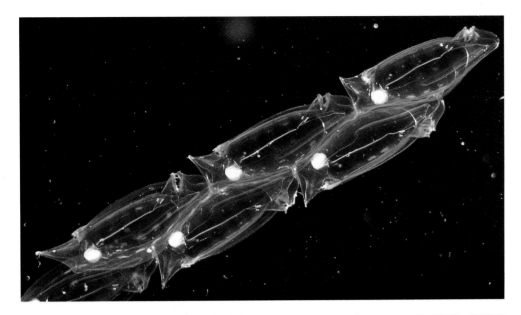

▲ **Verging on the vertebrate condition,** a lancelet. Here *Branchiostoma lanceolatum* lies with its tail in the gravel and its head protruding to maintain a respiratory and filter-feeding current.

◄ **Body plans of** (1) a sea squirt and (2) its "tadpole" larva, and (3) a lancelet.

are swept into the endostyle. In this ciliated gutter trapped food is collected up, bound into mucus strands, and passed into the mid-gut. The filtered water passes out into the atrium and leaves the body via the atrial pore. The midgut has a blind diverticulum leading forward. Backward the midgut leads to the intestine and then eventually to the anus.

The anatomy of these coelomates is complex in comparison with most invertebrates and there are some unusual features to it. The excretory system consists of sac-like nephridia lying above the gill bars. Each lancelet nephridium has a number of flame cells reminiscent of those seen in the flatworms (see p62), annelids and mollusks. In evolutionary terms the nephridia (which do not occur in other primitive chordates) are a far cry from the vertebrate kidneys, which must have evolved by a different route.

The muscle blocks are arranged in segments (myotomes) along either side of the body. A hollow dorsal nerve cord runs the length of the animal and shows very little anterior specialization or brain. As it passes back along the body it branches to supply muscles and other organs.

There is a simple circulatory system with blood vessels passing through the gills to collect oxygen and distribute it to the body. The sexes are separate. In *Branchiostoma* the gonads are arranged on both sides of the body at the bases of the muscle blocks, but in *Asymmetron*, whose name speaks for itself, they lie only on the right side. Sperms and eggs are released into the atrium by rupture of the gonad wall. They pass out to the sea through the atriopore. Fertilization occurs in the open water and a swimming larva or tornaria develops. This lives a dual life for several months as it feeds and matures. In the daytime it lies on the seabed but when darkness comes it swims and joins the plankton. When it has attained a length of about 0.2in (5mm) it metamorphoses and becomes bottom-dwelling and more sedentary. The tornaria is the chief distributive phase.

Adult lancelets live in shallow water inshore and occur from temperate to tropical regions. Although they are really burrowers, inhabiting sands and gravels, they can emerge from their burrows and swim actively for short periods. Because of its characters *Branchiostoma* provides an important lesson in our understanding of vertebrate organization, and its simple structure makes it quite easy to identify the basic chordate features, many of which are seen in humans themselves.   AC

vertebrates, and the front end of the animal, although quite distinct, lacks the well-developed brain, eyes and other sense organs as well as the jaws associated with vertebrates.

In species of *Branchiostoma*, a hood extends over the mouth equipped with slender, tentacle-like cirri instead of jaws. These form a sieve which assists in rejecting particles of food too large for the lancelet's suspension-feeding habit. The oral hood leads to the extensive pharynx via a thin flap of tissue, the velum. The pharyngeal wall is composed of many gill bars and it is the action of the cilia situated on those bars that draws the water and suspended food particles into the body by way of the oral hood. The gill bars act as a filter for food as well as providing a surface for the absorption of oxygen. When water passes through the gill bars, food particles are trapped and passed down to the floor of the pharynx, where they

BIBLIOGRAPHY

The following list of titles indicates key reference works used in the preparation of this volume and those recommended for further reading.

General Invertebrate Biology

Alexander, R. McN. (1979) *The Invertebrates*, Cambridge University Press, Cambridge.

Barnes, R. D. (1982) *Invertebrate Zoology*, 4th edn, Holt-Saunders, Philadelphia.

Barrington, E. J. W. (1982) *Invertebrate Structure and Function*, Van Nostrand Reinhold, New York.

Clark, R. B. (1964) *Dynamics in Metazoan Evolution*, Oxford University Press, Oxford.

Fretter, V. and Graham, A. (1976) *A Functional Anatomy of Invertebrates*, Academic Press, London, New York, San Francisco.

Grzimek, B. (ed) (1972) *Grzimek's Animal Life Encyclopedia* vols 1 and 3, Van Nostrand Reinhold, New York.

Hyman, L. H. (1940–67) *The Invertebrates*, vols 1–6, McGraw-Hill, New York.

Marshall, A. J. and Williams, W. D. (1982) *Textbook of Zoology: Invertebrates*, Macmillan, London.

Meglitsch, P. A. (1972) *Invertebrate Zoology*, 2nd edn, Oxford University Press, Oxford.

Moore, R. C., Lalicker, C. G. and Fischer, A. G. (1952) *Invertebrate Fossils*, McGraw-Hill, New York.

Russell-Hunter, W. D. (1979) *A Life of Invertebrates*, Macmillan, New York. Collier Macmillan, London.

Invertebrate Groups

Berquist, P. R. (1978) *Sponges*, Hutchinson, London.

Bliss, D. E. (ed) (1982–) *The Biology of the Crustacea*, vols 1–10, Academic Press, London and New York.

Berrill, N. J. (1950) *The Tunicata, with an Account of the British Species*, Ray Society, London.

Burton, M. (1963) *Revision of Classification of Calcereous Sponges*, British Museum (Natural History), London.

Cheng, T. C. (1973) *General Parasitology*, Academic Press, New York.

Crofton, H. D. (1966) *Nematodes*, Hutchinson, London.

Dales, R. P. (1967) *Annelids* (2nd edn), Hutchinson, London.

Darwin, C. R. (1851–54) *A Monograph of the Sub-class Cirripedia*, 2 vols, Ray Society, London.

Fretter, V. and Graham, A. (1962) *British Prosobranch Molluscs*, Ray Society, London.

Gibson, R. (1972) *Nemerteans*, Hutchinson, London.

Ingle, R. (1980) *British Crabs*, British Museum (Natural History), London, and Oxford University Press.

King, P. E. (1973) *Pycnogonids*, Hutchinson, London.

Lapage, G. (1963) *Animals Parasitic in Man*, Dover, New York.

Lincoln, R. H. (1979) *British Marine Amphipoda: Grammaridea*, British Museum (Natural History), London.

Manton, S. M. (1977) *The Arthropoda: Habits, Functional Morphology and Evolution*, Clarendon Press, Oxford.

Morton, J. E. (1968) *Molluscs*, Hutchinson, London.

Nichols, D. (1969) *Echinoderms* (4th edn), Hutchinson, London.

Ramazzotti, G. and Maucci, W. (1983) *Il Phylum Tardigrada* (3rd edn), Memorie dell Istituto Italiano di Idrobiologia Dott. Marco de Marchi, vol 41.

Robin, B., Pétron, C. and Rives. C. (n.d.) *Les coraux de Nouvelle-Calédonie, Tahiti, Réunion, Antilles*, Editions du Pacifique, Tahiti.

Rudwick, M. J. S. (1970) *Living and Fossil Brachiopods*, Hutchinson, London.

Ryland, J. S. (1970) *Bryozoans*, Hutchinson, London.

Schumacher, H. (1976) *Korallenriffe*, BLV Vorlagsgesellschaft, Munich, Berne, Vienna.

Stephenson, T. A. and Stephenson, A. (1928 and 1935) *The British Sea Anemones*, 2 vols, Ray Society, London.

Tattersall, W. M. and Tattersall, O. M. (1951) *The British Mysidacea*, Ray Society, London.

Thompson, T. E. (1976) *Biology of the Opisthobranch Molluscs 1*, Ray Society, London.

Thompson, T. E. and Brown, G. H. (1984) *Biology of the Opisthobranch Molluscs 2*, Ray Society, London.

Veron, C., Pichon, M., Wijsman-Best, M. and Wallace, C. C. (1976–84) *Scleractinia of Eastern Australia*, parts I–V, Australian Institute of Marine Sciences Monograph Series.

Warner, G. F. (1977) *The Biology of Crabs*, Elek, London.

Yonge, C. M. and Thompson, T. E. (1976) *Living Marine Molluscs*, Collins, London.

Regional Invertebrate Field Guides and Identification

Barratt, J. and Yonge, C. M. (1972) *Collins Pocket Guide to the Seashore*, Collins, London.

Campbell, A. C. (1984) *The Country Life Guide to the Sea Shores and Shallow Seas of Britain and Europe*, Country Life Books, London.

Campbell, A. C. (1982) *The Hamlyn Guide to the Flora and Fauna of the Mediterranean Sea*, Hamlyn, London.

Conseil Permanent International pour l'Exploration de la Mer, *Fiches d'Identification du Zooplankton*, Charlottenlund Slot, Denmark. (A series of papers on plankton identification.)

Dakin, W. J. (1952) *Australian Seashores*, Angus & Robertson, Sydney and London.

Day, J. H. (1974) *Marine Life on South African Shores*, Balkema, Rotterdam.

Linnean Society of London, and the Estuarine and Brackish Water Sciences Association, *Synopses of the British Fauna* (new series), E. J. Brill, Leiden. A periodical dealing with identification and general bionomics of many invertebrate groups.

Riedel, R. (1963) *Fauna and Flora der Adria*, Paul Perey, Hamburg and Berlin.

Rickets, E. F. and Calvin, J. (1960) *Between Pacific Tides* (3rd edn), Stanford University Press, Palo Alto.

GLOSSARY

Abdomen a group of up to 10 similar segments, situated behind the THORAX of crustaceans and insects, which in the former group may possess appendages.

Acoelomate having no COELOM (main body cavity).

Acrorhagi groups of NEMATOCYSTS.

Americ having a body not divided into SEGMENTS.

Ampulla a small contractile fluid reservoir associated with the tube-feet of some echinoderms.

Anisogamous of reproduction, involving gametes of the same species that are unalike in size or form.

Antennae the first pair of head appendages of uniramians and the second pair of crustaceans.

Antennules the first pair of head appendages of crustaceans.

Apophysis an outgrowth or process on an organ or bone.

Atrium the volume enclosed by the tentacles of an endoproct; also the chamber through which the water current passes before leaving the body of a sea squirt or lancelet.

Axon a long process of a nerve cell, normally conducting impulses away from the nerve cell body.

Axopodium a stiff filament or pseudopodium which radiates outward from the body of a heliozoan or radiolarian.

Benthic associated with the bottom of seas or lakes.

Binary fission a form of asexual reproduction of a cell in which the nucleus divides, and then the CYTOPLASM divides into two approximately equal parts.

Biomass a measure of the abundance of a life form in terms of its mass.

Bipectinate comb-like, of structures with two branches, particularly the OSPHRADIUM and/or CTENIDIUM in some mollusks.

Biramous of those arthropods (eg crustaceans) with forked ("two-branched") appendages.

Bothria long, narrow grooves of weak muscularity found in Pseudophyllidea (an order of tapeworm).

Branchial hearts contractile hearts near the base of each CTENIDIUM in certain mollusks.

Brood sac a thoracic pouch of certain crustaceans into which fertilized eggs are deposited and where they develop.

Buccal mass a muscular structure surrounding the RADULA, horny jaw and ODONTOPHORE of a mollusk (not a bivalve).

Budding a form of asexual reproduction in which a new individual develops as a direct growth from the body of the parent.

Calcareous composed of, or containing, calcium carbonate as in the spicules of certain sponges or the shells of mollusks.

Caudal relating to the tail or to the rearmost SEGMENT of an invertebrate.

Cecum a blindly-ending branch of the gut or other hollow organ.

Cephalothorax the fusion of head and anterior thoracic segments in certain crustaceans to form a single body region

which may be covered by a protective carapace.

Cerata (sing. ceras) projections on the back of some shell-less sea slugs, often brightly colored, which may bear NEMATOCYSTS from cnidarians and may act as secondary respiratory organs.

Cercaria a swimming larval form of flukes; produced asexually by REDIA larvae while parasitic in snails; cercaria infects a new final or intermediate host via food or the skin.

Chaetae the chitinous bristles characteristic of annelid worms.

Chela the pincer-like tip of limbs in some arthropods.

Chelicera one of the first pair of appendages behind the mouth of a chelicerate.

Chelicerate a member of the phylum Chelicerata.

Chitin a complex nitrogen-containing polysaccharide which forms a material of considerable mechanical strength and resistance to chemicals; forms the external "shell" or cuticle of arthropods.

Chloroplast a small granule (plastid) found in cells and containing the green pigment chlorophyll, site of photosynthesis.

Chordate a member or characteristic of the phylum Chordata, animals which possess a NOTOCHORD.

Chromatophore a cell with pigment in its CYTOPLASM.

Cilia (sing. cilium) the only differences between FLAGELLA and cilia is in the formers' greater length and the greater number of cilia found on a cell; flagella measure up to $22\mu m$, cilia up to $10\mu m$.

Ciliary feeding feeding by filtering minute organisms from a current of water drawn through or toward the animal by CILIA.

Ciliated having a number of CILIA on a surface which beat in a coordinated rhythm; ciliary action is a common method of moving fluids within an animal body or over body surfaces, employed in CILIARY FEEDING, and a common means of locomotion in microscopic and small animals. The ciliated ciliates are a major class (Ciliata) of protozoans.

Cirri (sing. cirrus) in barnacles, paired thoracic feeding appendages; in protozoans, short, spine-like projections in tufts called CILIA; in annelids, broad flattened projections situated dorsally on segments; in flukes and some turbellarian flatworms the cirrus is the male copulatory apparatus.

Coelom the main body cavity of many TRIPLOBLASTIC animals, situated in the middle layer of cells or MESODERM and lined by EPITHELIUM. In many organisms the coelom contains the internal organs of the body and plays an important part in collecting excretions which are removed via NEPHRIDIA or coelomoducts.

Coelomocyte a free cell in the COELOM of some invertebrates which appears to be involved with the excretion of waste material, wound healing and regeneration.

Conjugation the union of gametes, or two cells (in certain bacteria); or the process of sexual reproduction in most ciliates.

Corona the characteristic, ciliated, wheel-like organ at the anterior end of rotifers.

Coxa the basal segment of an arthropod appendage which joins the limb to the body.

Cryptobiosis a form of suspended animation enabling an organism to survive adverse environmental conditions.

Ctenidium one of the pair of gills within the MANTLE CAVITY of some mollusks.

Cytoplasm all the living matter of a cell excluding the nucleus.

Detritivore an animal that feeds on dead or decaying organic matter.

Deuterostome a member of a major branch of multicellular animals (the others are PROTOSTOMES). The mouth is formed as a secondary opening and the original embryonic blastopore becomes the anus. The embryo undergoes radial cleavage, the body cavity (ENTEROCOELOM) arises as a pouch from the ENDODERM, and the central nervous system is dorsal.

Dextral of spirally coiled gastropod shells in which, as usual, the whorls rise to the right and the aperture is on the right where the shell is viewed from the side.

Diatom a single-celled alga, a component of the PHYTOPLANKTON.

Dioecious having separate sexes.

Diploblastic a multicellular animal having a body composed of two distinct cellular layers, the ECTODERM and ENDODERM.

Dinoflagellate a unicellular member of the PHYTOPLANKTON characterized by the possession of two FLAGELLA, one directed posteriorly, the other lying at right angles to the posterior flagellum.

Ectoderm (is) the superficial or outer germ layer of a multicellular embryo which mainly develops into the skin, nervous tissue and excretory organs.

Ectoparasite a parasite which lives on the outside of its host and may be permanently attached or only come into contact with the host when feeding or reproducing.

Endocuticle the inner layer of the crustacean CUTICLE, composed of CHITIN.

Endoderm (is) the innermost of the three germ layers in the early embryo of most animals, developing into, for example, the lining of the ENTERON or, in jellyfishes the lining of the ENTERON or, in many animals, the gut lining.

Endoparasite a parasite which lives permanently within its host's tissues (except for some reproductive or larval stages). Often there are PRIMARY and SECONDARY HOSTS for different stages of the life cycle. Endoparasites are typically highly specialized.

Endoskeleton an internal skeleton, as in echinoderms and vertebrates.

Enterocoelom a COELOM that is thought to have arisen from cavities in the sacs of the MESODERM of the embryo.

Enteron the body cavity of cnidarians which is lined with ENDODERM and opens to the exterior via a single opening, the mouth.

Epibenthic living on the seabed between lowwater mark and some 656ft (200m) depth.

Epicuticle the outer layer of the crustacean cuticle, a thin, non-chitinous protective layer.

Epidermis the outer tissue layer of the EPITHELIUM.

Epithelium a sheet or tube of cells lining cavities and vessels and covering exposed body surfaces.

Epizoic a sedentary animal which is attached to the exterior of another animal but is not parasitic.

Exoskeleton the skeleton covering the outside of the body, or situated in the skin.

Fibril a small fiber, or subdivision of a fiber; used as contractile ORGANELLES in protozoans.

Filamentous a type of structure, eg a crustacean gill, in which the branches are thread-like, but not sub-branched, and are arranged in several series along the central axis.

Filter feeding a form of suspension feeding in which food particles are extracted from the surrounding water by filtering. Filtering requires the setting up of a water current usually by means of CILIA, with mucus being used to trap particles and sometimes to filter them from the surrounding water.

Fission see BINARY FISSION.

Flagellum (plural flagella) a fine, long thread, moving in a lashing or undulating fashion, projecting from a cell.

Flame cell the hollow, cup-shaped cell lying at the inner end of a protonephridium, important in the excretory system of some invertebrates. The inner end bears FLAGELLA whose beating causes body fluids to enter the NEPHRIDIUM.

Fragmentation a form of sexual reproduction in which an organism produces eggs in SEGMENTS of its body, which then break off and themselves split after leaving the host body, allowing the eggs to develop eventually into new organisms.

Funnel part of the molluskan "foot" in cephalopods responsible for respiratory currents to the CTENIDIA and for jet propulsion.

Gametocyte a cell which undergoes MEIOSIS to form gametes; an oocyte forms an ovum (female gamete) and a spermatocyte forms a spermatozoan (male gamete).

Gastrozooid a type of individual POLYP in colonial hydrozoans which captures and ingests prey.

Generalist an animal not specialized, not adapted to a particular niche; may be found in a variety of habitats.

Gill the respiratory organ of aquatic animals.

Gill book a type of gill, possessed by eg horseshoe crabs, formed by the five posterior pairs of appendages on the OPISTHOSOMA.

Gonoduct the duct through which sperm and eggs are released into the surrounding water.

Hemal system a tubular system of undecided function found in echinoderms.

Hemocoel the major secondary body cavity of arthropods and mollusks which is filled with blood. Unlike the COELOM it does not communicate with the exterior and does not contain germ cells. However, body organs lie within or are suspended in the hemocoel. It functions in the transport and storage of many essential materials.

Hermatypic of corals, reef-building corals with commensal zooanthellae.

Holoplankton organisms in which the whole life cycle is spent in the PLANKTON.

Host see INTERMEDIATE HOST; PRIMARY HOST; SECONDARY HOST.

Hydrostatic skeleton a fluid-filled cavity enclosed by a body wall which acts as a skeleton against which the muscles can act.

Hyperparasite (verb hyperparasitize) an organism which is a parasite upon another parasite.

Infusariiform a larval stage of mesozoans produced in members of the order Dicyemida by the hermaphrodite RHOMBOGEN generation, and in the order Orthonectida by free-living males and females and reinfecting the host.

Intermediate host an organism which plays host to parasitic larvae before they mature sexually in the final or definitive host.

Interstitial living in the spaces between SUBSTRATE particles.

Isogamy (adjective isogamous) a condition in which the gametes produced by a species are similar, ie not differentiated into male and female.

Keratin a tough, fibrous protein rich in sulfur: the outer layer of the cuticle of nematode worms is keratinized.

Kinety in ciliate protozoans, a row of kinetosomes and FIBRILS; from kinetosomes arise CILIA, the fibrils linking each kinetosome in a longitudinal row.

Lacuna a minute space, in invertebrate tissue containing fluid.

Lymph an intercellular body fluid drained by lymph vessels; contains all the constituents of blood plasma except the protein, and varying numbers of cells.

Macronucleus one of two nuclei found in ciliate protozoans.

Macrophagous diet a diet of pieces which are large relative to the size of animal; feeding usually occurs at intervals.

Madreporite a delicate, perforated sieve plate through which seawater may be drawn into the WATER VASCULAR SYSTEM of echinoderms; may be internal (eg sea cucumbers) or prominent external convex disk (starfishes).

Malpighian tubule/gland a tubular excretory gland which opens into the front of the hindgut of insects, arachnids, myriapods and water bears.

Mantle a fold of skin covering all or part of the body of mollusks; its outer edge secretes the shell.

Mantle cavity the cavity between the body and MANTLE of a mollusk, containing the feeding and/or respiratory organs.

Maxilla paired head appendages of crustaceans and uniramians which are located behind the mandibles on the fifth segment. They act as accessory feeding appendages.

Maxilliped the first one, two or three pairs of thoracic limbs of malocostracan crustaceans which have turned forward and become adapted as accessory feeding appendages rather than being involved in locomotion.

Maxillule paired head appendages of crustaceans and uniramians which are located on segment six behind the MAXILLAE. They also function in the manipulation of food.

Medusa the free-swimming sexual stage of the cnidarian life cycle, produced by the asexual BUDDING POLYPS.

Megalopa a postlarval stage of brachyuran crustaceans in which, unlike the adult (eg crab), the abdomen is large, unflexed and bears the full number of appendages.

Meiosis cell division whereby the DNA complement is halved in the daughter cells. Compare MITOSIS.

Meroplankton organisms passing part of their life cycle in the PLANKTON, usually the larval forms of BENTHIC animals.

Merozoite a stage in the life cycle of some parasitic protozoans which enters red blood corpuscles of the host.

Mesoderm the cell layer of TRIPLOBLASTIC animals that develops into tissue lying between the ENDODERM and ECTODERM.

Mesogloea the layer of jelly-like material between the ECTODERM and ENDODERM of cnidarians (jellyfishes etc).

Metameric having many similar SEGMENTS constituting the body.

Metazoan an animal, as in the vast majority of invertebrates, whose body consists of many cells in contrast to protozoans which are unicellular; a member of the subkingdom Metazoa.

Microfilaria the larval form of filaroid nematode worm parasites found in the SECONDARY HOST, usually mosquitoes.

Microflora microscopic bacteria occurring in the soil.

Microhabitat the particular parts of the habitat that are encountered by an individual in the course of its activities.

Micronucleus one of two nuclei found in the protozoan ciliates, the smaller micronucleus provides the gametes during CONJUGATION. See MACRONUCLEUS.

Microphagous diet a diet of pieces of food which are minute relative to the animal's own size; feeding occurs continually.

Microtubule a very small long, hollow cylindrical vessel conveying liquids within a cell.

Mitosis cell division in which daughter cells replicate exactly the chromosome pattern of the parent cell, unlike MEIOSIS.

Molt periodic shedding of the arthropod EXOSKELETON. Possession of a hardened exoskeleton prevents continuous growth until the adult stage is reached. Molting occurs under hormonal control after the secretion of a new and larger CUTICLE. An increase in size occurs during the short period prior to the hardening of the new cuticle, involving water or air uptake into the internal spaces. New tissue then grows into these spaces after the hardening of the new cuticle, ie between molts.

Monoblastic organisms having a single cell layer (eg sponges).

Morphology the structure and shape of an organism.

Myotome a block of muscle, one of a series along the body of a lancelet, sea squirt larva, or vertebrate.

Nauplius the first larval stage of some crustaceans which is divided into three segments each possessing a pair of jointed limbs that develop into the adult's two pairs of antennae and the mandibles. The nauplius uses its limbs in feeding and locomotion.

Nekton aquatic organisms, such as fish, which, unlike the smaller PLANKTON, can maintain their position in the water column and move against local currents.

Nematocyst the characteristic stinging ORGANELLE of cnidarians (eg jellyfishes) located particularly on the tentacles. A short process at one end of the ovoid cell (cnidoblast) containing the nematocyst acts as a trigger opening the lid-like OPERCULUM. Water entering the cnidoblast swells the nematocyst, a long thread-like tube coiled up inside. The nematocyst discharges, ensnaring prey in its barbed coils or releasing poison down the tube into the victim.

Nematogen the first and subsequent early generations of certain mesozoans (order Dicyemida), parasites of immature cephalopods.

Nephridiopore the pore by which a NEPHRIDIUM opens to the exterior.

Nephridium an excretory tubule opening to the exterior via a pore (nephridiopore). The inner end of the tubule may be blind, ending in FLAME CELLS or it may open into the COELOM via a ciliated funnel.

Notochord a row of vacuolated cells forming a skeletal rod lying lengthwise between the central nervous system and gut of all CHORDATES.

Odontophore structure in the mouths of most mollusks, except the bivalves, that supports the RADULA.

Oligomericous having a few segments constituting the body.

Operculate the condition of gastropods having an OPERCULUM.

Operculum a lid-like structure; the calcareous plate on the top surface of the foot of some gastropods, serving to close the aperture when the animal withdraws into the shell.

Opisthosoma the posterior body region of chelicerates which may be segmented in primitive forms but generally has the segments fused.

Organelle a persistent structure forming part of a cell, with a specialized function within it analagous to an organ within the whole organism.

Osmotic pressure and regulation osmotic pressure is the force that tends to move water in an osmotic system, ie the pressure exerted by a more concentrated solution on one of a lower concentration. The body fluids of a freshwater animal will exert an osmotic pressure on the surrounding aqueous medium, causing water to enter the animal. Osmoregulation is the maintenance of the internal body fluids at a different osmotic pressure from that of the external aqueous environment.

Osphradium a patch of sensory EPITHELIUM located on gill membranes of mollusks.

Palp an appendage, usually near the mouth, which may be sensory, used as an aid in feeding, or in locomotion.

Papilla a small protuberance ("little nipple") above a surface.

Parapodium one of a pair of appendages extending from the sides of the segments of polychaete worms.

Pathogen an agent which causes disease, always parasitic.

Pedicellariae minute, pincer-like grooming and defensive structures on the body of sea urchins and starfishes.

Pedipalp an appendage borne on the third prosomal segment of chelicerates, sensory or prehensile in horseshoe crabs, adapted for seizing prey in scorpions, and sensory or used by the male in reproduction in spiders.

Peduncle a narrow part supporting a longer part, eg the muscular stalk by which the body of an endoproct is attached to the SUBSTRATE.

Pelagic of organisms or life-styles in the water column, as opposed to the bottom SUBSTRATE.

Pentamerism the fivefold RADIAL SYMMETRY typical of echinoderms.

Pericardial cavity the cavity within the body containing the heart. In vertebrates a hemocoelic space, which is an expanded part of the blood system, supplying blood to the heart.

Peristalsis rhythmic waves of contraction passing along tubular organs, particularly the gut, produced by a layer of smooth muscle.

Pharynx part of the alimentary tract or gut behind the mouth, often muscular.

Pheromone a chemical substance which when released by an animal influences the behavior or development of other individuals of the same species.

Phyletic concerning evolutionary descent.

Phylogeny the evolutionary history or ancestry of a group of organisms.

Phytoplankton microscopic algae that are suspended in surface waters of seas and lakes where there is sufficient light for photosynthesis to take place.

Pinnate of tentacles, GILLS, resembling a feather or compound leaf in structure, with similar parts arranged on either side of a central axis.

Pinnule a jointed appendage found in large numbers on the arms of crinoids giving a feather-like appearance, hence the name feather star.

Plankton drifting or swimming animals and plants, many minute or microscopic, which live freely in the water and are borne by water currents due to their limited powers of locomotion.

Planula the free-swimming ciliated larva of cnidarians (jellyfishes and allies).

Plasmodium the asexual stage of orthonectid mesozoans, resembling the protozoan plasmodium, which divide repeatedly by FISSION, filling the hosts' tissue spaces.

Pneumostome the aperture to the lung-like MANTLE CAVITY of pulmonates.

Polyp the stage, the most important in the life cycle of most cnidarians, in which the body is typically tubular or cylindrical, the oral end bearing the mouth and tentacles and the opposite end being attached to the SUBSTRATE.

Polysaccharide a carbohydrate produced by a combination of many simple sugar or monosaccharide molecules, eg starch and cellulose.

Primary host the main host of a parasite in which the adult parasite or the sexually mature form is present.

Proglottides (sing. proglottis) the segments which make up the "body" of a tapeworm. When mature, each proglottis will contain at least one set of reproductive organs.

Prokaryote cell having, or organism made of cells having, genetic material in the form of simple filaments of DNA, not separated from the CYTOPLASM by a nuclear membrane. Bacteria and blue-green algae have cells of this type.

Prosoma the anterior body region of CHELICERATES, comprised of eight SEGMENTS analagous to the head and THORAX of other arthropods, or the CEPHALOTHORAX of chelicerates. The segments are generally fused and are only distinguishable in the embryo.

Prostomium the anterior non-segmental region of annelid worms, bearing the eyes, ANTENNAE and a pair of PALPS; comparable to the head of other phyla.

Protoconch the first shell of a gastropod which is laid down by the larva.

Protonephridium a type of excretory organ in which the tubule usually ends in a FLAME CELL.

Pseudocoel the secondary body cavity of roundworms, rotifers, gastrotrichs and endoprocts, between an inner MESODERM layer of the body wall and the ENDODERM of the gut, ie it is not a true COELOM.

Pseudopodium (plural: -podia) a temporary projection of the cell when the fluid endoplasm flows forward inside the stiffer ectoplasm. Occurs during locomotion and feeding.

Pseudotrachea a branched TUBULE resulting from intuckings of the cuticle of certain terrestrial isopods which acts as a specialized respiratory surface. Pseudotracheae resemble the tracheae of uniramians and certain arachnids, although they have evolved independently.

Pygidium the terminal, non-segmental region of some invertebrates which bears the anus.

Radial symmetry a form of symmetry in which the body consists of a central axis around which similar parts are symmetrically arranged.

Radula the "toothed tongue" of mollusks, a horny strip with ridges or "teeth" on its surface which rasp food. Absent in members of the class Bivalvia.

Ray a radial division of an echinoderm, eg a starfish "arm".

Redia a larval type produced asexually by a previous larval stage of flukes (Trematoda). Lives parasitically in snails and reproduces asexually, giving rise to more rediae or to CERCARIAE.

Reticulopodium a type of PSEUDOPODIUM characteristic of the foraminiferans; reticulopodia are thread-like, branched and interconnected.

Rhombogen an hermaphrodite form of dicyenid mesozoan derived from a NEMATOGEN when the cephalopod host has reached maturity. It resembles the nematogen morphologically and gives rise to INFUSARIIFORM larvae.

Rostrum the anterior plate of the crustacean carapace, present in malacostracans, which extends toward the head and ends in a point.

Sclerotization hardening of the arthropod cuticle by TANNING.

Scolex the head region of a tapeworm, which attaches to the wall of the host's gut by suckers and/or hooks.

Secondary host the host in which the larval or resting stages of a parasite are present.

Sedentary sedentary organisms, or stages in the life cycle of certain organisms, permanently attached to a SUBSTRATE; as opposed to free living.

Segment a repeating unit of the body which has a structure fundamentally similar to other segments, although certain segments may be grouped together into TAGMATA to perform certain functions, as in the head, THORAX or ABDOMEN.

Septum a portion dividing a tissue or organ into a number of compartments.

Seta (plural setae) a bristle-like projection on the invertebrate EPIDERMIS.

Siliceous composed of or containing silicate, as in the skeleton of glass sponges.

Sinistral of gastropod shells, with whorls rising to the left and not as usual to the right (compare DEXTRAL).

Sinus a space or cavity in an animal's body.

Siphon a tube through which water enters and/or leaves a cavity within the body of an animal, eg in mollusks and in sea squirts.

Speciation the origin of species, the diverging of two like organisms into different forms resulting in new species.

Spermatophore a package of sperm produced by males, usually of species in which fertilization is internal but does not involve direct copulation.

Spermatotheca an organ, usually one of a pair, in a female or hermaphrodite that receives and stores sperm from the male.

Spicule a mineral secretion (calcium carbonate or silica) of sponges which forms part of the skeleton of most species and whose structure is of importance in sponge classification.

Spiral cleavage a form of embryonic division which occurs in PROTOSTOMES; in the spiral arrangement of cells any one cell is located between the two cells above or below it. In all other many-celled animals, ie DEUTEROSTOMES, there is radial cleavage.

Spore a single- or multi-celled reproductive body that becomes detached from its parent and gives rise directly or indirectly to a new individual.

Sporocyst a sac-like body formed by the miracidium larva of a blood fluke while within the snail, the intermediate host; produces numerous CERCARIAE, over 3,000 per day from a single sporocyst.

Sporozoite SPORE produced in certain protozoans which then develops into gametes.

Statocyst the balancing organ of a number of invertebrates consisting of a vesicle containing granules of sand or calcium carbonate. These granules move within the vesicle and stimulate sensory cells as the animal moves, so providing information on its position in relation to gravity.

Stolon the tubular structure of colonial cnidarian POLYPS that anchors them to the SUBSTRATE and from which the polyps arise.

Stomodeum a region of unfolding ECTODERM from which derive the mouth cavity and foregut in many invertebrates.

Strobila the "body" of a tapeworm, consisting of a string of segments, through which food is absorbed from the gut of the host.

Stylet a small, sharp appendage, for example in water bears, used to pierce plant cells.

Substrate the surface or sediment on or in which an organism lives.

Suspension feeding a feeding process in which small organisms and other matter suspended in the water are removed and consumed.

Symbiosis a close and mutually beneficial relationship between individuals of two species.

Synapse the site at which one nerve cell is connected to another.

Tagmata (sing. tagma) functional body regions of arthropods and annelids consisting of a number of SEGMENTS; eg the head, THORAX and ABDOMEN of crustaceans.

Tanning hardening of the arthropod cuticle achieved by the cross-linking of the protein chains by arthoquinones, involving also polyphenol and polyphenoloxidase catalysts.

Tegumental gland a gland below the EPIDERMIS of the crustacean cuticle. Ducts from the glands convey the constituents of the EPICUTICLE to the cuticle surface during molting, when the new epicuticle is formed.

Telson the posterior segment of the arthropod abdomen which is present only embryonically in insects. In certain crustaceans the telson is flattened to form a tail fin which is used in swimming.

Test an external covering or "shell" of an invertebrate, especially sea squirts (tunicates), sea urchins etc; it is in fact an internal skeleton just below the EPIDERMIS.

Thorax the segmented body region of insects and crustaceans which lies behind the head and which typically bears locomotory appendages. Up to 11 segments are present in crustaceans but only 3 in insects.

Torsion the process of twisting of the body in the larval stage of gastropods.

Trichocyst rod-like or oval ORGANELLE in the ECTODERM of protozoans which may discharge a long thread on contact with prey.

Triploblastic having three growing layers.

Trochophore an oval or pear-shaped, free-swimming, planktonic larval form of organisms from different phyla, including segmented worms and mollusks.

Tube-feet or podia—hollow, extensive appendages of echinoderms connected to the WATER VASCULAR SYSTEM that may have suckers, or serve as stilt-like limbs or be ciliated to waft food particles toward the mouth.

Tubule long hollow cylinder within a cell, normally for conveying or holding liquids.

Tunicin, tunic a form of cellulose, the main constituent in the fibrous matrix forming the tunic or test of sea squirts.

Uropod flattened extension of the sixth abdominal appendage of malacostracan crustaceans which together with the flattened TELSON form a tail fin used in swimming.

Vacuole a fluid-filled space within the CYTOPLASM of a cell, bounded by a membrane.

Veliger a free-swimming larval form of mollusks possessing a VELUM; develops from a TROCHOPHORE; foot, MANTLE, shell and other adult organs are present.

Velum the veil-like ciliated lobe of the VELIGER larva, used in swimming; also the inward-projecting margin of the umbrella in most hydrozoan MEDUSAE.

Ventral situated at, or related to, the lower bottom side or surface.

Vermiform a worm-like larval stage of dicyenid mesozoans formed within the axial cells of the NEMATOGEN generation or, generally, worm-like.

Water vascular system or ambulacral system, a system of canals and appendages of the body wall that is unique to echinoderms, derived from the COELOM and used, eg, in locomotion in starfishes.

Zoea a planktonic larval form of some decapod crustaceans which possesses a segmented THORAX, a carapace and at least three pairs of BIRAMOUS thoracic appendages. In contrast to the antennal propulsion of the NAUPLIUS these thoracic appendages are used in locomotion. The abdominal pleopods appear but are not functional until the postlarval stage.

Zoochlorella a symbiotic green alga of the Chlorophyceae which occurs in the amoeboicytes of certain freshwater sponges, the gastrodermal cells of some hydra species and the jelly-like connective tissue (parenchyme) of certain turbellarian flatworms.

Zooid a member of a colony of animals which are joined together; may be specialized for certain functions.

Zoospore a motile spore which swims by means of a FLAGELLUM, is produced by some unicellular animals and algae, and is a means of asexual reproduction.

Zooplankton small or minute animals that live freely in the water column of seas and lakes and consists of adult PELAGIC animals or the larval forms of pelagic and some BENTHIC animals; most are motile, but the water movements determine their position in the water column.

Zygote a fertilized ovum before it undergoes cleavage.

INDEX

Picture Acknowledgments

Key *t* top, *b* bottom, *c* center *l* left, *r* right

Abbreviations: A Ardea. AN Agence Nature. ANT Australasian Nature Transparencies. BCL Bruce Coleman Ltd. NHPA Natural History Photographic Agency. NSP Natural Science Photographs. OSF Oxford Scientific Films. PEP Planet Earth Pictures. SPL Science Photo Library.

Cover BCL/Neville Coleman. 1 A/Ron and Valerie Taylor. 2–3 BCL/Frans Lanting. 4–5 A/Wardene Weisser. 6–7 BCL/Bob and Clara Calhoun. 8–9 PEP/Laurence Gould. 10, 11 OSF. 13*t* A/V. Taylor. 13*b* SPL. 14*t* OSF/D. Allan. 14*b* NHPA/B. Wood. 15 PEP/J. Mackinnon. 16 C. Howson. 16–17 PEP/K. Lucas. 18 Biophoto Associates. 19, 20*b* OSF/P. Parks. 20–21 NHPA/M. Walker. 22–23 OSF/P. Parks. 26 Biophoto Associates. 27 SPL. 29 Biophoto Associates. 30–31 B. Picton. 31*b* NSP/I. Bennett & F. Myers.

32 OSF. 33 C. Howson. 34–35 NHPA/B. Wood. 39 PEP/B. Wood. 40–41 NHPA/B. Wood. 42–43 PEP/J. Greenfield. 44*b* BCL. 44–45 OSF/F. Ehrenström. 46 PEP/L. Madin. 47 OSF/P. Parks. 49 Biophoto Associates. 50–51 J. Walsh. 52*b* OSF/P. Parks. 52–53 B. Picton. 54 OSF/J. Cooke. 55 C. Howson. 56–57 OSF/S. Foote. 61 B. Picton. 62–63 Premaphotos. 63*b* OSF/P. Parks. 64*tl* S. Stammers 64*r* SPL/C. Ellis. 66, 67 SPL. 68–69 Biofotos/H. Angel. 69*b* D. Weathered. 70 Premaphotos. 71 OSF/A. Kuiter. 72*t* ANT/R. & D. Keller. 72*b* SPL/M. Dohrn. 76 A/P. Morris. 76–77 OSF/R. Kuiter. 80 BCL. 83 OSF/K. Atkinson. 84–85 OSF. 87 NHPA/G. Bernard. 90–91 PEP/P. David. 92–93 OSF/D. Shale. 94–95 BCL. 94*b* NHPA/N. Callow. 95*c* AN/Chaumeton. 95*b* Premaphotos. 98 A/V. Taylor. 99, 100 BCL. 100–101 PEP/D. Maitland. 102–103 BCL. 104*t* NHPA/J. Carmichael. 104*b* PEP/D. Maitland. 105

BCL. 106, 106–107 NSP/I. Bennett. 108 Biofotos/H. Angel. 109 BCL. 110–111 SPL. 111*b* OSF/P. Parks. 112*b* C. Howson. 112–113 PEP/B. Wood. 116 NHPA/J. Carmichael. 116–117 OSF/G. Bernard. 117*b* Biophoto Associates. 120–121 OSF/G. Bernard. 120*b* Biofotos/H. Angel. 121*b* B. Picton. 122 PEP/D. Maitland. 123 BCL. 124–125 NSP/I. Bennett. 125*t* PEP/J. Lythgoe. 126 OSF/R. Kuiter. 127*t* OSF/G. Bernard. 127*b* M. Fogden. 128–129 A/V. Taylor. 130*t* PEP/C. Pétron. 130*b* Biofotos/S. Summerhays. 131 PEP/F. Jackson. 133 Biofotos/H. Angel. 134–135 PEP/K. Lucas. 136*b* B. Picton. 136–137 BCL. 137 PEP/C. Pétron. 138*b*, 138–139 NHPA/A. Bannister. 139*b* PEP/B. Wood. 142 NHPA/B. Wood. 143 C. Howson. 144–145 PEP/A. Svoboda. 146–147 OSF/G. Bernard. 147*b* B. Picton. 148–149 J. Jamieson. 150 OSF/D. Shale. 151*t* OSF/ P. Parks. 151*b* OSF/G. Bernard.

Artwork

Abbreviations BC Barbara Cooper. ML Mick Loates. RG Roger Gorringe. RL Richard Lewington. SD Simon Driver.

1 2*t*, 1 2*b*, 1 2*r* SD. 16–17 RL. 18 BC. 23*t*, 23*b* SD. 24–25 RG. 27, 28, 29, 33, 34, 35 SD. 36–37 RG. 38, 42, 47, 48–49, 51, 53, 54, 55, 56 SD. 58–59 RG. 60, 62, 65*t*, 65*b*, 66, 67, 68, 70 SD. 72–73 RL. 73*b* SD. 74–75 RG. 76 SD. 76–77 RL. 79*t* SD. 79*b* RG. 80–81, 81, 82, 86 SD. 88–89 ML. 90 SD. 91 RL. 96–97 ML. 98–99 RL. 101 SD. 108, 110 RL. 114–115 ML. 116 SD. 118–119 R. 126, 131 SD. 132–133 ML. 134 SD. 140–141 RG. 142 SD. 143, 146 RL. 150 SD.